Front

Tao
of
Life
The Fractal Gift

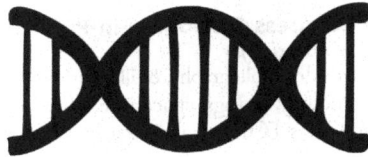

Katya Walter, Ph.D.

-:≡:-

**The author thanks the
Institute for Neuroscience and Consciousness Studies
of Austin, Texas
for its helpful support and encouragement**

-:≡:-

THIS BOOK IS
VOLUME 3, FOURTH EDITION
IN THE
TOUCHING GOD'S TOE SERIES

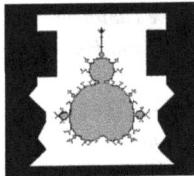

TO VISIT THE DOUBLE BUBBLE UNIVERSE, GO TO......
https://www.katyawalter.com

TO VISIT KATYA WALTER'S YOUTUBE CHANNEL, GO TO...
KatyaWalterYouTube

Kairos Center Publications
Box 142086
Austin, Texas 78714
kairospublications@gmail.com

Tao of Life: The Fractal Gift by Katya Walter, Ph.D.
Volume 3 in the *Touching God's TOE series*, 4th edition
Copyright © 2004 by Kairos Center; 4th edition by Kairos Center
Editor: Jennie Rosenblum   Art by Adele Aldridge, Adrian Frye, & Katya Walter
      - or from Wiki Commons or Creative Commons
Paperback 4th edition published 2019   ISBN 978-1-884178-52-8
Electronic 4th edition published 2019   ISBN 978-1-884178-77-1

Library of Congress Cataloging-in-Publication Data
Walter, Katya - *Tao of Life: The Fractal Gift*
Includes table of contents, appendix, bibliography, & illustrations
  1. Physics—gravity, cosmology, strings, spacetime, dimensions, fractal topology
  2. Religion—Touching God's TOE, spirit & science, religions, divine love
  3. Gravity—gravitation, unification, forces, emergent properties
  4. Chaos Theory—Lorenz attractor, fractals, chaos patterning, complexity
  5. Philosophy—Plato, Taoism, Chinese thought
  6. Mathematics—nonlinear, analinear, fractals, analog & linear number
  7. Mysticism—mystic love, remote viewing, dreams, I Ching, synchronicity systems
  8. Title: *Tao of Life: The Fractal Gift*

-:::-

Katya Walter's Books
Chaosforschung (in German) - *1992 - Diederichs Verlag*
Dream Mail: Secret Letters for your Soul - 1995 - Kairos Center
Tao of Chaos: Merging East and West - 1994 - Kairos Center - This original book was split and augmented to become Volumes 2 and 3 of the *Touching God's TOE series*, first published in 2004, and updated in a 4th edition, as shown below.

Touching God's TOE series, 4th Edition

Table of Contents

Initial German Review

The German publication of a initial single volume sparked this series. It was published in German as **Chaosforschung** *by Diederichs Verlag. Claus Claussen wrote this review that appeared in the magazine* **Neues Denken und Handeln** *in November 1992. That original book was later split and amplified into Volumes 2 and 3 of this series, so this translated review appears at the front of both volumes. [Permission was given to adapt the next-to-last paragraph slightly to fit the larger scope of the whole series.]*

"*Universal Life Pattern* could be a subtitle for this lofty theme that will pique your interest in the Orient. It might also be called *Breaking a Universal Code*, because it opens the door on a fascinating view of life. Number, more exactly, archetypal number, is the key to this research on chaos theory, Chinese philosophy, and DNA.

"Katya Walter, prominent philosopher from Texas, a Ph.D. who also has studied at the Jung Institute in Zurich and taught for a year at Jinan University of Guangzhou, goes to the source of life's dynamic pattern in her book. She describes how the DNA spiral of our linear-minded Western science relates to the analog-style thinking of the old I Ching. She shows that the genetic code and I Ching function through the same chaos patterns, and that the physical system of DNA can be translated mathematically into the psychic system of the I Ching.

"Other scientists, and especially Martin Schönberger (1973) in his book *Verborgener Schlüssel zum Leben—Weltformel I Ging im genetischen Code*, have earlier pointed out an astonishing correspondence between the genetic code and the I Ching. Walter makes reference to this work, but adds a new analog perspective, even enlightenment beyond Schönberger's book, going deeper and wider. Very concretely and beyond speculation, she lays bare a decodable correlation between amino acids and hexagrams. She shows that biochemical laws and old wisdom are connected through this mathematical pattern. It garbs old Eastern truth in new Western clothing. This chaos supersystem is provable with new terminology and computer graphics.

"Threading through the awesome labyrinth of this stunning theme, your guide Katya Walter continually startles you back into clarity with her personal engagement in the search for truth. She gives sidelong glances into her dreams, talks of her experiences and frustrations, and even jokes along the path. At such times her tone, normally scientific and yet crisp with a refreshing simplicity, takes on a more poetic lilt. The author takes an informative stroll through the chaos garden as she explores its profound central theme, approaching it from three distinct vistas: I Ching, chaos theory, and genetic code. This sight-seeing tour is designed to render each path fascinating yet familiar. Otherwise the waves of scientific proof could become too big.

"Above all, this carefully crafted work is a treasure trove chock full of jewels. Finally, there is a special paradoxical treasure at the bottom of the chest: without ever leaving the groundwork of science, it moves beyond logic into universal values."

> The universe is like a safe
> to which there is a combination.
> But the combination is locked up
> in the safe.
>
> *Peter De Vries*

Introduction–What Is this Book?

From the Author: To understand the I Ching better, during 1990-91, I taught at Jinan University in China for a year and studied the I Ching with scholar Zhang Luanling. There I also had many informative discussions with scholar Tan Shi-lin. I did this because I was exploring the puzzle of why the genetic code and the ancient I Ching show so many parallels.

To a casual eye, the genetic code seems very different from the I Ching. One is physical; the other is philosophical. One is new knowledge; the other is very old. DNA codes for organic matter, but the I Ching claims to show the flow of universal mind, which ancient China called the Tao. The two systems appear to be very different concepts existing in unrelated areas of study explored by historically disparate cultures situated far apart in both space and time.

My puzzle was this: both systems, DNA and I Ching, seem to echo each other in both math and meaning! But why? Slowly I found the reason. The genetic code's double helix and the I Ching's math are two variants of a shared construct. They are both based on the same underlying co-chaos paradigm.

I first explored this idea in *Tao of Chaos*, published in English in 1994. It eventually became the precursor for this series you're now reading called *Touching God's TOE*—whose very title indicates that physics and metaphysics can merge into a shared cosmology, a notion first proposed in recorded history by Plato.

The first book in this series, *Double Bubble Universe*, gives an overview of the paradigm, explaining how cosmologies in physics and metaphysics can merge into a Theory of Everything or TOE. There I also describe an extraordinary dream that initiated the idea. The second book, *Co-Chaos Patterns*, discusses the I Ching's deceptively simple math and how it is a uniquely stable and trustworthy version of nonlinear dynamics that I call analinear. There I also describe the wandering path of breadcrumbs that I followed on the trail of writing this series.

My purpose in Books 1 and 2 was to acquaint the reader thoroughly with basic concepts in the TOE, laying a foundation so that here in Book 3, I can show you how the I Ching and genetic code are based on the same paradigm. Since this book holds tedious coding descriptions, I also include some beautiful fractal images made by using that same code in its genetic and I Ching layouts.

FOR AN OVERVIEW OF THE WHOLE SERIES, LOOK FOR THE SERIES SUMMARY AT THE BACK.

I am carefully laying out for you a trail of breadcrumbs in this series of books. It offers you multiple layers of evidence showing that both known codes, genetic and I Ching, are fractal variants of a deeper master code of co-chaos whose polarized information holds the Double Bubble universe together.

All three codes use a polarized pair of pairs: the genetic code's four base molecules; the I Ching's four bigrams; the master code's four primals (space, time, matter, and energy). All three codes develop polarized triplets that pair-bond into 64 polarized 6-packs. Together, all three codes offer us a Rosetta Stone with two known codes that can reveal and unlock the obscure master code.

You'll find this book is arranged a bit differently from the others in this series. The first six chapters continue to alternate between odd-numbered chapters that are scientific and even-numbered chapters that are more philosophical and personal. However, those details of cross-coding the genetic code and I Ching can get tedious. For fun and relief, I added a playful interlude in Chapters 8, 9, and 10, where the hidden number structures underlying this co-chaos paradigm are turned into artful patterns in vivid colors.

After that, Chapters 11 and 12 show parallels between the 55 atoms of DNA's base molecules and the 55 dots of ancient China's He Tu map. Chapter 13 describes the *Mawangdui Silk* I Ching found in China in 1973. As usual, the last chapter discusses a specific hexagram. In Volume 3, it is Hexagram 3.

-:::-

From the Editor: This book is Volume 3 in the dazzling *Touching God's TOE* series, 4th edition. In this volume, Katya Walter, Ph.D., shows how correlating the genetic code with ancient China's I Ching provides a Rosetta Stone to decode the master code that generated our universe. She also explores some of its philosophical aspects, including a huge, unified mind that exists in nature itself, accessible via dreams, remote viewing, and other altered states.

This series began as one volume, *Chaosforschung*, published in German in 1992, then as *Tao of Chaos* published in English in 1994. That book was later split and amplified into Volumes 2 and 3 of this series, *Touching God's TOE*.

Volume 3 has 14 chapters in 112 sections. It includes a *Series Summary*, *Bibliography*, and *Reviews*, along with 104 listed images, graphics, and charts. The color ebook version has an interactive table of contents and 86 e-links that act as informative footnotes. Its text is completely searchable and receives electronic updates. It is also hand-edited to hold color graphics that allow greater distinctions in images and charts. Consider getting both the print and ebook versions of this book for a greater range of information and versatility.

Each book in this series has its own symbol. The symbol for Volume 3 is the DNA helix from *Science Icons* by Woodcutter. You see it here.

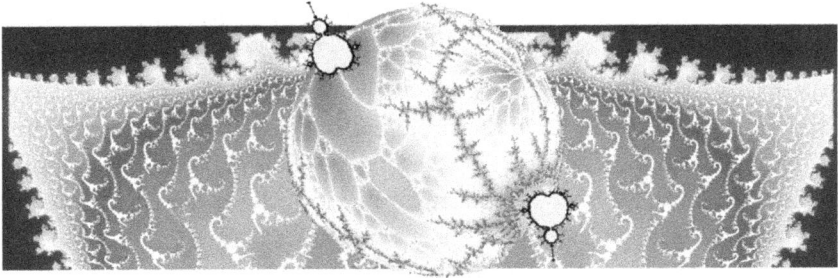

Chapter 1: DNA Plan & RNA Builder

1. DNA holds the building plan

The genetic code is a dazzler that defies entropy. This time traveler carries the plan of organic life from the past and evolves it onward into the future. It codes for all life—warm-blooded, cold-blooded, or bloodless. Virus. Amoeba. Twig. Moth. Raccoon. Fish. Human.

DNA holds the plan for all cellular species. Every cell in your body starts with DNA. Its plan records how your cells should grow, function, and reproduce to deliver your heritage into the future. RNA is tasked with turning DNA's plan into amino acids, then into proteins that produce iterations of each species as unique individuals. For instance, it gives one human baby brown eyes, dark hair, and skill with tools, while another gets green eyes, red hair, and a gift for gab—and those two very different people may be siblings in the same family.

In 1953, James Watson and Francis Crick proved the genetic code is carried on a double helix. One helix combines a circle's repetition with a line's directional thrust. DNA's double helix bonds *two* lines of that directional thrust via molecular rungs that turn it into a twisty ladder. Its rungs are made of polarized molecules organized as 3-packs (codons) along both spirals, which then pair-bond into polarized 6-packs of data. Some have compared DNA to a meshed zipper...not in designer jeans, mind you, but in designer genes.

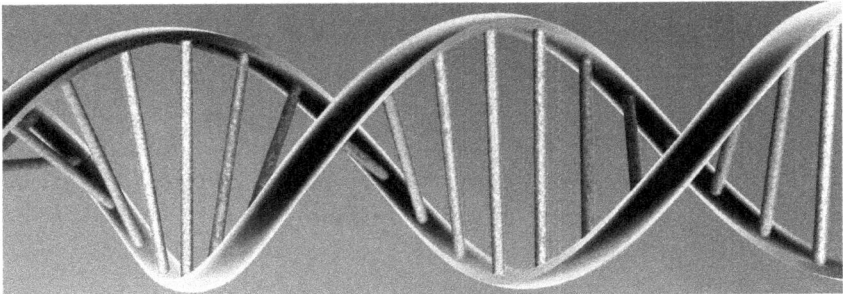

DNA's double helix

How does the DNA "zipper" of the double helix develop? Inside a cell are many copies of the four base molecules: T, A, C, and G. They are a polarized pair of pairs...and they *could* act like two couples do-si-do-ing in a square dance. But to make the double helix, they instead form a long, spiraling double-line dance. The polarized molecular 3-packs on each spiral pair-bond into rung-like 6-packs that link together both spirals. For clarity, the double helix below is differentiated into a black strand and a white strand.

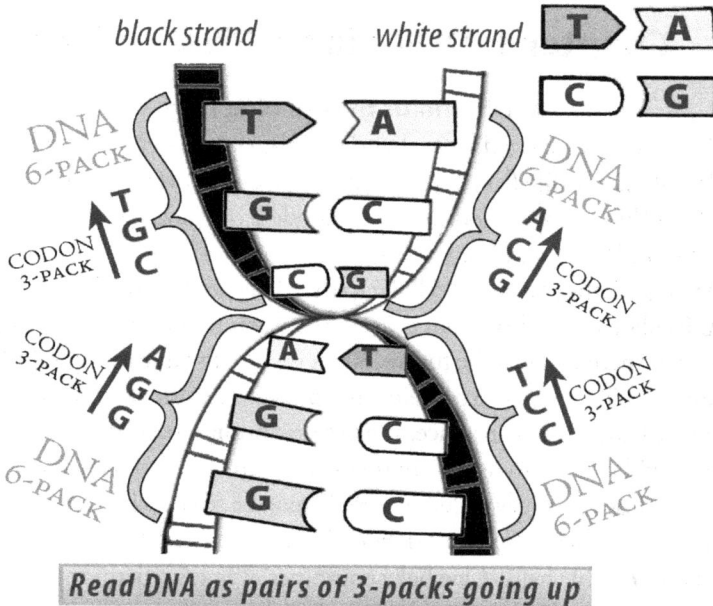

Read DNA as pairs of 3-packs going up

Double helix with a DNA 6-pack in brackets

Along both strands, molecules of T, A, G, and C line up. Each arrow describes a molecular 3-pack called a codon. The codons pair-bond along the double helix into 6-packs of molecular data. Brackets on this graphic indicate two different DNA 6-packs. Each 6-pack holds two codons. Each molecule in a codon is polarized to attract its opposite polarity on the other strand. Thus, one strand's polarized sequence automatically defines the opposite strand's polarized sequence.

These polarized, cross-referenced bonds stabilize the double helix and lock it into a nearly fail-safe record of genetic information. Its double-entry bookkeeping ensure DNA's genetic security to safeguard the inherited data of all cellular life. It protects the blueprint for every egg, sperm, and cell.

2. RNA was born to roam

DNA holds the plan to build organic life, but an organism cannot appear unless that plan gets built. A double helix must unzip its two strands to start the building process, allowing the sturdy, fail-safe DNA to expose its two vulnerable strands of molecules. That unzipping is necessary to actualize the plan.

When the double helix unzips, each of the two solitary DNA strands becomes a parental template. Its polarity will attract specific free-floating molecules in the cellular soup, and along its exposed length appears a new strand of RNA molecules. Below right, you see a codon on the white parental DNA strand attracting three free molecules to form a codon on the gray RNA strand, left. Notice that RNA's 3-pack mimics (almost) the absent black parental strand.

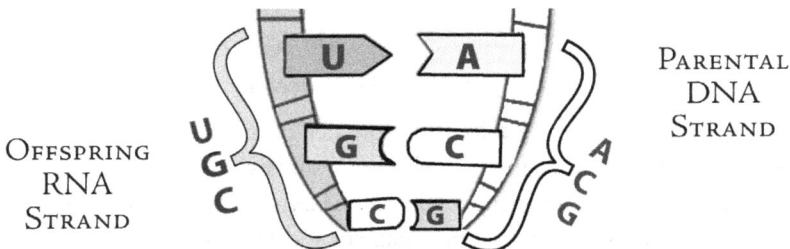

A gray RNA offspring strand develops along the white DNA parental strand

The gray offspring mimics the code of the absent black parental strand... but not with exact fidelity. Whenever it tries to attract a floating T (Thymine), it instead attracts a variant molecule U (Uracil). Thus any T molecule that existed on the black strand will instead be a U molecule on the gray RNA offspring strand. This means that T occurs only in DNA, and U occurs only in RNA.

That slight molecular difference alters the fate of the RNA offspring strand. Why? Well, U has one less carbon atom and two fewer hydrogen atoms than T did, so on the gray RNA strand, its smaller U molecules form weaker bonds than heftier T molecules managed to do for the parental DNA double helix.

U's slightly different, smaller molecule decrees that RNA's gray offspring strand cannot bond into a stable double helix like DNA's parental black and white strands managed to form. Instead, the U makes such a fragile bond that its lack of grip turns RNA into a solitary wanderer. It can never pair-bond.

RNA was born to roam. Below, left, contrast the offspring RNA's unbonded 3-pack containing U to, right, parental DNA's bonded 6-pack containing T.

OFFSPRING RNA 3-PACK   PARENTAL DNA 6-PACK

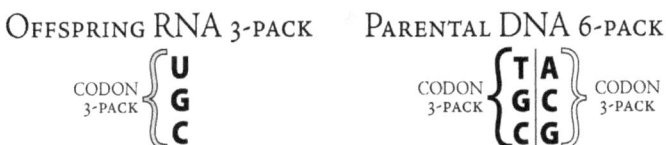

RNA offspring 3-pack & DNA parental 6-pack

3. RNA roams off to build the plan

Due to those weaker U molecules, the offspring must detach from the parental DNA template and go to work. As an independent strand, it wanders off in the cellular soup holding a copy of molecular data that replicates the black DNA strand, but not exactly, since all the parental T's have now become U's.

This wandering gray strand is messenger RNA (often called mRNA), and it was vital in making a new kind of vaccine that activates the innate immune system against the Covid-19 virus.

Here is part of the genetic code message on this mRNA string...

·..A A G A C U C G A U G A C U A G ..·

String of molecules to decode

How will the plan coded by this sequence of molecules on the mRNA' strand get read and turned into proteins that build living cells? Olivier Gascuel and Antoine Danchin put it this way in *Data Analysis Using a Learning Program*: "Using an analogy between a living organism and a building, one may say that the DNA represents the architect's plans and proteins represent the building materials and machinery." As the RNA strand roams in the cell's liquid, a ribosome comes up and attaches to the strand to decipher its coded message. The simplified ribosome below is amusingly known as the two-blob model.

Ribosome decoder

It looks like a snail, but think of this ribosome as a tiny protein-making factory, full of workers with names like rRNA and tRNA. In the next graphic, a ribosome factory chugs from left to right along the string of code written in molecules.

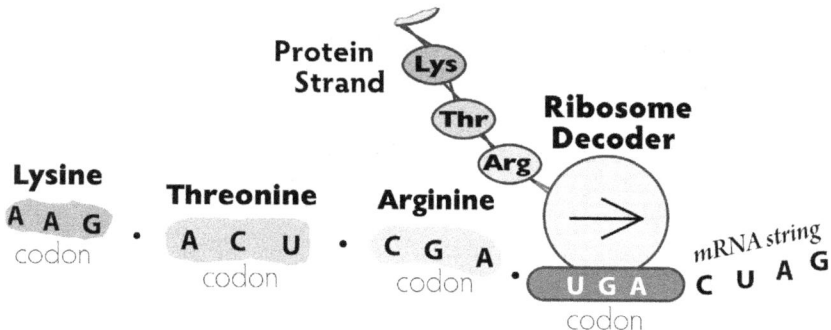

Ribosome decoder reading the mRNA string

The ribosome factory workers scan the long message on the mRNA strand and "chunk" it into molecular triplets. Codons. Each codon is a 3-pack of molecular data that gets decoded as an amino acid that's then translated into a protein. The path of this little portable ribosome factory is like a conveyer belt decoding mRNA's data into that strand of proteins you see rising above it.

Proteins of materialized code literally write the story of every organism into existence. Scientists even call this act *writing*. They say a split DNA strand *transcribes* its message on the forming RNA offspring, then releases it to roam until a ribosome attaches and *translates* that message into proteins.

This particular little ribosome factory, however, is just about ready to stop work. How do we know? Its lower chamber is deciphering a traffic codon right now—UGA—and UGA is the traffic codon that signals *Full Stop*. It commands: "Message ends here, so stop!"

There are four traffic codons—just one *Start* command, but three different *Stop* commands. They're all tasked to tell the ribosome when to start and stop working on an RNA strand. Sometimes two different *Stop* codons will appear in succession, like amber and red traffic lights. Then the genetic code is signaling to the ribosome, "Caution, get ready to stop. Now…really stop!"

4. Subcodes & overlay coding

The genetic code is very clever. For instance, like a spy, it can overlay one coded message atop another in an offset manner. Each subcode handles a specific part of the full message. Why would nature bother to layer its codes? Because it is such an efficient way to save space, time, material, and energy.

The easiest technique is breathtakingly simple: a ribosome can decode a string of molecules as three different messages, depending upon where it starts to "chunk" the string into 3-packs (codons). Here is part of a string… **AAGACUCGAUGACUAG**…that can be decoded in three different ways, depending on where the ribosome begins to chunk the RNA message into codons—

As…**AAG—ACU—CGA—UGA—CUA—G**…
Or as…**A—AGA—CUC—GAU—GAC—UAG**…
Or as…**AA—GAC—UCG—AUG—ACU—AG**…

A ribosome can start at the first molecule on the string, chunk into 3-packs, and read it as codons for certain amino acids. Then it can begin again at the second molecule and read the same string anew as different codons that stand for other amino acids. It can even recommence at a third location to decode yet a different set of amino acids. So the message is altered by merely moving its start-point for decoding just one position up or down the string! This actually happens in certain circumstances of the genetic decoding process. Clever!

According to Brian Hayes, senior writer for *American Scientist*, "What fascinated me about the code-breaking effort was how quickly a biochemical puzzle…was reduced to an abstract problem in symbol manipulation. Within a few months, all the messy molecular complexities were swept away, and the goal was understood to be a mathematical mapping between messages in…different alphabets. The methods for devising codes came from combinatorics; the proposed solutions were judged largely by the criteria of information theory."

The co-chaos paradigm underlying the genetic code and the I Ching math figures will let us do a mapping between their messages in different alphabets. However, to accomplish this, we must use a series of steps.

This book will show you how the 64 hexagram 6-packs of the I Ching can code for the 64 molecular 6-packs of DNA. Each hexagram can shorthand two codons of DNA. But a hexagram can also shorthand a single 3-pack of RNA, due to a fractal property in both codes that lets variants be embedded at different scales. Moreover, the hexagram's philosophical dynamic will even fit its amino acid task! Is that even possible? Indeed it is! The two systems really do parallel each other, both mathematically and philosophically.

In the more technical, odd-numbered chapters, we'll reduce this to an abstract problem in symbol manipulation that maps between the genetic code and I Ching math. Moreover, we'll also parallel their dynamics with their philosophies (what followers of Plato called metaphysics.) Cross-coding both systems shows they are two different ways to express the same underlying paradigm.

The main points to remember: DNA's double helix holds the building plan for organisms. The rungs of its twisty ladder are made of molecules. They sit in polarized 3-packs on both spirals, and they pair-bond across both spirals into 6-packs that hold DNA's cross-referenced plan safe inside the double helix.

But DNA's double helix can unzip into two separate strands. Each strand is a parental template. Its polarity attracts molecules to form an offspring RNA strand that mimics the absent parental strand, but substituting U for T…and U's weaker bonds release the RNA strand to wander with its message. Ribosomes come up to decode its message; they attach to it, scan it, chunk it, and turn it into proteins.

This overview lists just a few basic genetic principles that I chose to spotlight certain dynamics relevant to the co-chaos paradigm. However, each subset of the genetic code has its own workers and tasks, with many specializations that allow its processes to perform various separate, intermediary steps to generate life. Recognizing some major dynamics in the genetic code will help us correlate them with the I Ching math figures, which in turn can help us spot and correlate similar parallels existing in the obscure master code.

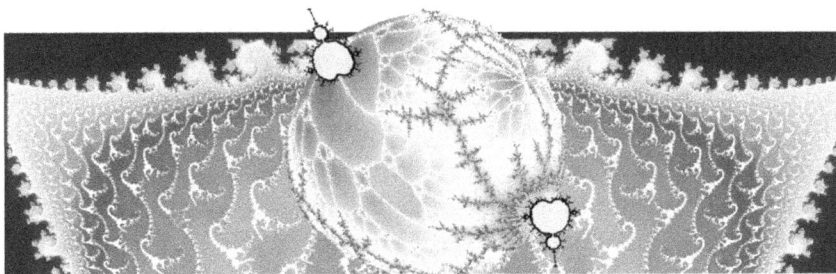

Chapter 2. Mixing Math and Mysticism

1. Exploring polarity—West & East

West and East took very different paths philosophically, but both sides of the globe explored the concept of polarity. Western civilization is rooted in Greco-Roman-Judean thought, with Socrates (470-399 BCE) as its foremost early philosopher. Socrates had a prize student: Plato (428-347 BCE). Plato wrote down Socrates' philosophy that the universe holds two polarized realms: Ideal and Real, which vie in perpetual struggle...abstract vs. concrete...mind vs. matter...form vs. contents. Western culture adopted this view that ongoing, polarized conflict churns out reality, with life a constant duel between polarities.

Plato likewise had a prize student: Aristotle (384–322 BCE). Aristotle took the scientific measure of things to establish empirical proofs. His approach fostered a divide-and-conquer attitude toward measuring and evaluating matter itself—divide and conquer that physical substance!

Aristotle also had a prize student, Alexander the Great (356–323 BCE). He set out to divide and conquer the known physical world. By the time of Julius Caesar (100-44 BCE), that divide-and-conquer mindset appeared in his boast: "*Veni, vidi, vici*—I came; I saw; I conquered." He ruled Rome, even named himself "dictator in perpetuity" for about a month until he was assassinated.

Seeing life as perpetual conflict promoted an *either-or* tendency in the West. In such a mindset, logic-chopping splits life into *either-or* oppositions; it's either *this* or *that*. Mind or matter. Black or white. Good or evil. Yes or no. You win, I lose...or the reverse. No matter which pole wins, it's still an *either-or* stance.

Despite the West's cultural allegiance to an *either-or, win-lose, profit-loss* attitude in life, it could not extinguish the rejected other pole, nor the promise of *both-and* union in a condition of wholeness. The holistic side of the brain simply does not go away, not even if the linear, logic-chopping side of the brain tries to repress that resonant mystery into the unconscious.

But Western culture might have gone very differently. Before Socrates, the philosopher Heraclitus (535-475 BCE) described the universe as a dynamic, emergent flow: "No man ever steps in the same river twice." He also spoke of transcending one's focus on either pole of a duality to perceive the larger unity beyond such division, favoring a transcendent third condition of wholeness.

Dualistic Western philosophy even tried to transcend the *either-or* dualism that gripped its thought by postulating logic's version of a third condition: the *tertium quid* in Aristotle's 3-step syllogism: "If this, and this, then this"... where logical sequencing of two terms brings forth a triumphant third result.

After 1,000 years of dogma debate, 13th century Christianity brought forth q mystical version of a third thing. The Gnostic view that Spirit was feminine became heresy, and Spirit was re-gendered to become male. It often appeared in paintings as an inseminating ray of light beaming down on human Mary to deliver Jesus into the center of a patriarchal, all-male Trinity as Father, Son, and Holy Ghost. In that way, the third male condition of Jesus became a triumphant, transcendental stairway to heaven, bridging that painful gap between God's immensity and the human condition's frailty. In the 18th century, Kant fretted over unsolvable polarities or *antinomies*. Fichte tweaked that dualism into the triad of *Thesis, Antithesis, Synthesis*, which he popularized.

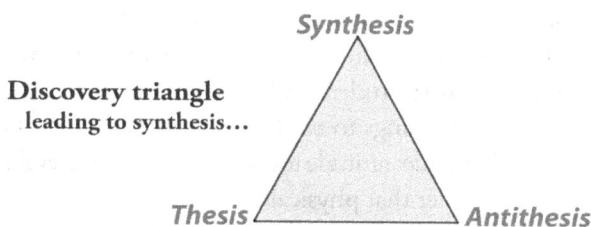

Discovery triangle
 leading to synthesis...

Synthesis

Thesis   *Antithesis*

The discovery triangle of thesis, antithesis, & synthesis

Yet none of this was nearly so complex a resolution of polarity as we find in Eastern thought. Ancient China originated the yin-yang concept. Legend says it was codified by emperor Fuxi (Fu Hsi in Wade-Giles transcription) around 3000 BCE. The polarity concept was elaborated and nuanced into the I Ching's 64 hexagram figures of co-chaos math, signifying a Taoist philosophy of balance that became known throughout the Far East.

The I Ching treats the universe as a unity whose polarities operate in an ongoing synergy that promotes change. This synergy is evident in the tai chi ball. Assertive, bright yang signals the emerging foreground; receptive, dark yin signals a shadowy background. White yang cannot emerge without black yin to hold it.

Tai Chi ball

Thus, unlike in Western thought, yin and yang polarity do not fight each other; instead, they enable each other to exist. Due to their continually shifting balance, the polarized partners also have the dynamic potential to change into each other, as hinted by the two dots of opposite color in the tai chi ball.

2. The dynamic flow of change

You experience this alternating principle at work in the *vase/faces* image of gestalt psychology, where black and white alternate as field and ground. If you stare at the image below, your eyes will automatically keep switching focus.

Vase/faces image of gestalt psychology

While your eyes are focusing on the white faces, now white is acting yang-dominant. When your eyes switch focus instead to the black vase, now black becomes yang-dominant, for it is what grabs your attention. If white really always acted yang, and black always acted yin, then you'd always notice only the white faces and never the black vase.

In the graphic below, move your eyes along the row of images. Yang-dominance shifts between *Vase* and *Faces* as your attention hits each image.

Yang	Yang	Yang	Yang	Yang	Yang	Yang
vase	*faces*	*vase*	*faces*	*vase*	*faces*	*vase*

Yin & yang in a shifting field-ground gestalt

Hermann Hakan noted this alternation when he said that the *vase/faces* image has two attractor points—*vase* and *faces*—and the viewing eye switches back and forth between them at an unpredictable rate. This repeated shift of focus between *Vase* and *Faces* throws successive attractor points into the two wells of possibility.

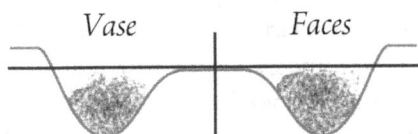

Vase   Faces

Bifurcation into two wells of possibility

Imagine this shifting dynamic of black and white in the tai chi ball as a series of successive film clips where the ball seems to keep turning.

Yin-yang dynamic seen as turning motion

In *From Clocks to Chaos*, Leon Glass and Michael Mackey present a diagram of the solutions of a differential equation in a period 2 limit cycle oscillation. They show that its vector forces actually sketch in the rotating tai chi symbol!

Vector forces in a period 2 limit cycle oscillation

Yang is finally just whatever stands out at the moment to grab your attention. Yin is the background that contains it. When your mind eventually realizes that both partners together create the whole—as modeled by the tai chi ball—then you achieve a larger view that encompasses its polarized dynamic.

The ancient Chinese saw yin and yang as two complementary conditions that rise to a higher unity transcending their polarity. First comes a state of undifferentiated unity. Things seem alike, just a gray blur ⬤. Slowly you begin to differentiate between this and that. The gray blur differentiates into ◑.

Ah, comes the dawn! What was formerly a blurry generalization has now gained a division between two states. The ancient Chinese likened this boundary to the separation of sky and earth. Bright, transparent sky symbolized yang, and dark, opaque earth symbolized yin. As complementary partners, each helped to define the other. The main idea is that two poles are necessary for movement, and development occurs only through change. The momentum of the tai chi ball's dynamic creates a continual alternation between polarities, for instance, as seen between day and night, good and bad, or happy and sad.

You'll find this kind of thought in the ancient Taoist story of the man whose horse ran away one day. The other villagers lamented and said, "Oh, your horse ran off! Too bad." But the man just shrugged and said, "Maybe yes, maybe no."

The next day his horse came wandering back home…with two wild horses following behind. The villagers rejoiced, saying, "Now you have three horses! How lucky you are!" But the man only shrugged and said, "Maybe yes, maybe no."

The next day his oldest son tried to tame one of the wild horses. He fell off and broke his leg. The villagers said, "Oh no! Now your oldest son, so big and strong, has a broken leg. He cannot help you in the field. How unlucky for you!" But the man shrugged and said, "Maybe yes, maybe no."

The next day the emperor's army came into the village, conscripting all able-bodied young men, but they could not be bothered to cart along the son with a broken leg...on and on it goes. A version of this same tale was told 3,000 years later in Alaska by Marilyn, Native American medical receptionist, in a television episode of *Northern Exposure* in 1994. Tales travel.

You no doubt have experienced for yourself how a piece of seemingly terrible luck can eventually bring real benefit. I Ching philosophy says no issue is merely binary. Of course, things may seem black or white at first: "Either it is, or it isn't. That's all there is to it!"...until the situation evolves.

Nothing remains simple, including the I Ching. The remarkable Chinese Kangxi emperor (his reign name means Strong Brilliance) held the throne for 61 years (1661-1722 CE). He said, "I have never tired of the Book of Changes [I Ching], and have used it for fortune-telling, and as a source of moral principles. The only thing you must not do, I told my court lecturers, is not to make this book appear simple, for there are meanings here that lie beyond words."

3. The momentum of life

The ancient Chinese said the I Ching algorithm taps into the Tao of universal mind, helping one look backward and forward in time to see a bigger picture. A strange thought, but no stranger than the actual genetic code that tracks backward and forward in time beyond your own body to encompass the bigger picture of your species. In that way of thinking, the genetic code is a time traveler. It spans the entire history of our species, and indeed, all life on Earth. It contains the past like a time capsule of our organic heritage, starting with the first living cell. Moreover, it looks toward the future of each species. Although individual members of each species die off continually, genes evolve to carry forward new iterations in new variants, promoting life's momentum.

Genes opt for the future. In each species, its genetic plan maintains and evolves what has gone before, working to develop a more diversified, adapted animal, insect, or plant. The history coded inside those genes maintains the living system, even while it constantly adapts to changing conditions. In this way, organic life maintains its base even as it also evolves. Remarkable!

This momentum evident in life itself has long puzzled scientists, for it appears to contradict the second law of thermodynamics, which says that all material in the universe breaks down due to spontaneous chemical and physical processes. Nicolas Sadi Carno, a French engineer and physicist, first formulated this idea well in 1824. He said if no energy enters or leaves a closed system, its entropy never decreases; instead, the closed system evolves toward thermal equilibrium, cooling down into a state of maximum entropy. Entropy itself was defined this way by Prof. Frank L. Lambert: "Entropy is simply a quantitative measure of what the second law of thermodynamics describes: the dispersal of energy in a process in our material world."

In the material world, things run down. Machines break. Fires die. Baths cool. Ice melts. Wood rots. Gases mingle. The second law of thermodynamics predicts that due to its dispersing energy, our universe's closed system will end in a *heat death* where all energies reach equilibrium and nothing changes anymore.

But is our universe truly a closed system? Physicists now speculate that many universes may exist in soap-bubble clusters, multiverse membranes, and so on. Further, does entropy even apply to the organic living systems that we find here on Earth? Physicist Erwin Schrodinger said in *What is Life?* (1951) that living systems appear to have an energizing trait that counters entropy. He called it *negative entropy*—which Leon Brillouin shortened to *negentropy*.

Life itself acts negentropic. Although individual units deteriorate and die, life itself keeps on evolving the many different species into branches and sub-branches that crowd each other for new windows of opportunity in the complex global ecology. We see the thrust of evolving life here on Earth that diversifies even as it also maintains its vital base. We humans appear to thrive on increasing complexity. By feeding on the energy of a star, bacteria, money, gasoline, and Eggs Florentine, our species survives and proliferates. So far.

But *why* does life act negentropic? Hans Driesch (1867–1941) hypothesized that some teleological cause gives life its vital force by directing the living system toward a goal. He called this theory *vitalism* and suggested that it works by the detailed timing of micro-processes that are "set free into actuality."

Driesch called this unknown kind of power *entelechy* and said it would not contradict the second law of thermodynamics if its non-spatial cause is merely enacted in space. The mind-bending quirkiness of quantum mechanics had not yet been discovered, so the Newtonian-style mindset then prevailing dismissed Driesch's vitalism as unsubstantiated woolly-mindedness.

Then during the 1960s, chemist and inventor James Lovelock studied the atmosphere on Mars for NASA and worked on detecting any life there. That work led him and microbiologist Lynn Margulis to develop in the 1970s the Gaia hypothesis that the Earth is a self-regulating system that tries to maintain the conditions for life. The Gaia theory proposed that our globe is a huge entity that lives, evolves, and responds to stimuli in a coherent and self-regulating manner. Lovelock and Margulis built a big data bank of biological evidence for the planet Earth as a complex system of interactions that can be considered a giant living organism. The scientific community at large has started to agree that this may indeed be so.

In 1972, Lila Gatlin said in *Information Theory and the Living System* that classical ideas of entropy are inadequate to explain the complexity of the living system. She suggested that the tools of mathematics may clarify this problem. But then she remarked wryly, "Mathematics is simply a way of expressing concepts that anyone can understand in a way that very few can understand."

Gatlin also said, "The genetic code is a small subroutine of a master program which directs the machinery of life. We have no idea what the language of this master program is like, but we can be sure that it has always evolved, is now evolving, and will continue to evolve in the future."

By 1981, Rupert Sheldrake said in *A New Science of Life* that since quantum physics arrived on the scene, the idea of a life force no longer appears so full of holes. He suggested that an unknown set of scientific variables may explain how the life force works, and he posited a "morphogenetic field" of gene-changing energy. He said until we know how the field works, any theory will be unsatisfactory, for that unknown field necessarily operates on the physical world from a realm of networking connectivity that we do not yet perceive.

Biologists now recognize that life somehow keeps generating and elaborating itself in ways that could not have been predicted from looking at the original first cell. Some even propose that organic evolution occurs at a far faster rate than could happen through mere random mutation, and thus a foresightful or "purposeful" mutation seems to be operating...although Ludwig von Bertalanffy, Gerhard Roth, Antti Revonsuo, and others dismiss it as mere "systems causality."

One version of this trait of purposeful mutation is *epigenetics* (from two Greek roots meaning *after* and *creation*). This term entered contemporary genetics vocabulary in the 1990s. Epigenetics looks at trait changes that occur when various factors switch genes on or off, thereby affecting how the cells read those genes. Such changes do not occur due to changes in the genetic code itself; rather, they are due to the environment that a gene finds itself in... and different environments can turn some genes on or off.

In *Metamagical Themas: Questing for the Essence of Mind and Pattern*, Douglas Hofstadter wondered at the ability of life to stabilize, reproduce, and complicate itself endlessly: "Ever since self-replicating molecules came about, they have been reproducing like mad and proliferating in ever more varieties.... We ourselves are huge self-replicating molecule-heaps. Ever upward builds this dizzying spire of self-replicating structures. What gives this whole movement any coherent direction? How and why does complexity evolve from simplicity?"

4. A universal mind in the universal body

How did the universe manage to cook us up? What recipe did it use? At first, its container of spacetime held nearly uniform contents of hot, plasmic, mattergy soup. As the soup slowly cooled, it evolved into more complex matter and energy structures—suns, planets, oceans, continents—locales fit to incubate complex molecular clusters of organisms, including us.

But how did the universe manage to do that? This TOE suggests it happened because the universe itself is alive. What if the universal body has a universal mind? What if its evolving mind slowly tweaked its evolving body into the stuff that supports us, that is us? This TOE suggests that universal mind is at work in every level of reality, in the tiny fraction we perceive and all the rest that we don't.

Ancient peoples have already long viewed the Earth as a living organism. Many so-called "primitive" cultures felt cherished by a tissue of invisible connection while living on the Earth's physical body. Catholic eco-theologian Thomas Berry pointed out that "Although this belief was never central to Western thought tradition, it maintained itself consistently on the borders of Western consciousness as the 'anima mundi' concept, the soul of the world."

During the Greco-Roman heyday, linear logic won out in the West over the many ancient mystery cults. That connective, holistic mindset was marginalized and funneled into a religion that became the most successful mystery cult of all time. Christianity honors a dead god who magically revived to live forever in each believer's heart, bringing faith, hope, and love to make one's logical life more satisfying by balancing it out with emotion. How wonderfully mysterious!

This TOE proposes that the living system of diverse organisms is not isolated to planet Earth. Life is part of something larger that accesses the heritage of the living universe itself, which went through many stages just to make possible the diverse little organisms scattered in it here and there like bacteria.

The universe has evolved all its factors of organization. It evolved the so-called inorganic stuff…elements merged into suns that acquired solar systems that swirled in galaxies that clumped into superclusters…as well as evolving the complex forms that we on Earth call organic life…where microbes proliferate and whales school, where ants nest and people lounge in Roman baths.

By saying this, I do not imply that our universe itself is God, the Grand Organizing Design. I merely suggest that the universal body has a universal mind. The ancient Chinese called it the Tao. India called it the Akasha. Westerners talk of the collective unconscious, the Over-Soul, higher mind. All point to something larger than this world, maybe this solar system or galaxy, perhaps even this universe. Is there something beyond? Does it matter?

5. Do math & mysticism mix?

In his Nobel speech, *Scientific and Religious Truths*, Werner Heisenberg pointed out that the much-quoted statement "God is a mathematician" derives from Plato. He said mathematicians confine their vision to mathematical proofs, but Plato himself did not: "Having pointed out with the utmost clarity the possibilities and limitations of precise language, he switched to the language of poetry, which evokes in the hearer images conveying understanding of an altogether different kind…the language of images and likenesses is probably the only way of approaching the 'one' from more general domains."

Thus, it is no wonder that physicists sometimes take poetic license by naming their discoveries in fanciful ways—the Strange and Charm quarks, for instance. Physicists sometimes also flirt with the numinous mystery of huge patterns suspended beyond ordinary logic. Isaac Newton (1642–1727) developed Newtonian mechanics, but he also owned 169 books on alchemy and 447 books on astrology, according to his library's inventory.

How can one merge math and mysticism effectively? Jesuit missionaries tried to do so in 17th and 18th century China. They began a *Figurist* movement based on the I Ching's mathematical shorthand of "hexagram figures" because "all mathematics seems to be rooted in the I Ching's permutations."

The Figurist Jesuits proposed that if the I Ching were considered in its "genuine purity," free from the many later additions bundled in over centuries that ballooned the short *Changes* into the extensive *Book of Changes*, the result would be identical to the laws of nature.

Jesuit Figurist Joachim Bouvet also encouraged polymath Gottfried Wilhelm Leibniz (1646–1716) to spend years of effort connecting Christian theology to the I Ching. Leibniz viewed the I Ching as a universal language combining math with a philosophy that could scientifically prove God's existence. However, Leibniz died before much came of it, which greatly relieved Pope Clement XI. He declared that Leibniz had gone astray theologically.

Meanwhile in China, Jesuit Figurists told the Kangxi emperor that the Bible was based on the same natural laws as those found in the I Ching. He was interested to hear their ideas, and his decree of 1692 said, "The Europeans are very quiet; they do not excite any disturbances in the provinces, they do no harm to anyone, they commit no crimes, and their doctrine has nothing in common with that of the false sects in the empire, nor has it any tendency to excite sedition….

"We decide therefore that all temples dedicated to the Lord of Heaven [meaning the Christian God], in whatever place they may be found, ought to be preserved, and that it may be permitted to all who wish to worship this

God to enter these temples, offer him incense, and perform the ceremonies practised according to ancient custom by the Christians. Therefore let no one henceforth offer them any opposition."

However, Pope Clement XI soon issued two papal bulls forbidding any Chinese Christian converts to perform a collateral worship of Confucius or their own ancestors. So the Kangxi emperor in return decreed that if the Catholic Church refused to let its Chinese converts continue performing collateral ceremonies according to ancient Chinese custom, then Westerners could no longer preach Christianity in China.

The Kangxi emperor said of the papal bulls, "I have concluded that the Westerners are petty indeed.... To judge from this proclamation, their religion is however, no different from other small, bigoted sects...."

It is notable that Eastern religions tend to allow a *both-and* syncretism of worship, while Western religions insist on *either-or* exclusion of worship.

Indeed, by 1773, Pope Clement XIV was suppressing Jesuits around the world as too innovative, declaring they threatened Catholic dogma. Why? He said the Jesuit Figurists had become irrational about numbers and were far too interested in integrating scientific and mystic thought.

History had already shown that getting irrational over numbers can be dangerous. In ancient Greece, Pythagoras (570-495 BCE) was a mathematician and philosopher whose followers believed reality could be understood in terms of whole, rational numbers. A rational number can be stated in an exact ratio... for example, ⅔. Then the Pythagoreans discovered a horrific fact: some numbers are irrational. An irrational number cannot be expressed as an exact ratio. For example, pi (π) has no exact number ratio. It is only *approximately* 22/7.

The geometer Pappus reported that the first Pythagorean who divulged the horrific secret that some numbers have no exact ratio was drowned. It has been variously described as a punishment by other members, by the gods, or even as a remorseful suicide. Hmm...imagine killing someone else or yourself upon discovering that irrational numbers exist. That horror of losing the known ratios in life is why we still sometimes call a frantic person "irrational."

6. Quantum organics

This TOE says our universe is alive. It is based on the numbers of co-chaos patterning. The master code's simplicity evolved the complexity of universal fractal patterns that manifest in space, time, matter, and energy. A lesser fractal variant is the genetic code, whose simplicity evolved all life forms on Earth. This TOE shows that the I Ching math provides yet another fractal variant, one whose symbols can shorthand the master code generating the universe itself.

Yet this universe is only one of many artful designs in the great catalog of universes. My dream showed me many universes, and they were just a part of the larger Grand Organizing Design. Sure, in mathematical terms, the whole is equal to the sum of its parts. But our universe's mathematics was put together in such a way that it created life, making it become more than the sum of its parts. As Aristotle pointed out in *Metaphysica,* we experience a holism beyond any sequencing of numbers, facts, data. He said "…the whole is not a mere heap. The totality is something beyond the parts. There is a cause of unity…."

My great dream showed that we tiny humans here on Earth are not mere lonely flames that spark up for a moment and die. Yes, each body dies, but not the higher awareness that some would call *soul,* a connective thread to the larger life beyond us. We tap into such insights sometimes in meditation, in prayer, in wonder, in dreams that tune us toward the ongoing source linking us all beyond each death. All of us contribute to the Grand Organizing Design.

This TOE even suggests that to clear up some of the mysteries of current cosmology—the puzzles of inflation, Bell's theorem, dark matter, dark energy, and the nature of black holes, for instance—we would do better to speak of quantum *organics* rather than quantum *mechanics.* Perhaps our universe forms the equivalent of God's pancreas…I don't know. In that dream where I met God, I saw that all of the universes together are less than half of God. I could only discern enough to understand that I could never know God fully. The Grand Organizing Design is too big, and I am too small. But what I did perceive convinced me that we are a known, evolving part of the Grand Organizing Design, and what we do with our lives matters. It affects the whole.

In the numbers that generate the constant emergence of our own universe's reality, evolution and entropy are complementary processes that point to one purpose: we are evolving the Grand Organizing Design. Since it is much more than all of the universes put together, for short, I just call it God.

7. Healing a polarized split by transcending both poles

As we transition toward a global culture, the electronic web of worldwide communication connects us, one to another, like a giant nervous system. To move past our simplistic wars of polarized conflict, we can look toward a larger unity beyond the many differences and disagreements. We can come to see our planet more and more tangibly as a living system, even as we also grow more aware of the ecological perils that we are bringing upon it, upon ourselves, so that we strive to become more than parasites leaching Earth's life away.

How to heal the polarized splits? By transcending polarity. Acknowledging the merits on both sides of an argument allows me to see what is good about both poles of an issue, transcending the oppositional conflicts that pull people willy-nilly, each demanding loyalty for only its side. Doing so allows me to move calmly past the binary split of absolute good and evil, saints and sinners, exorcists and demons, white hats and black hats, heroes and villains.

Oh, look! Now we see the bigger picture that offers more than just two choices. This gives us more options to handle life's difficulties instead of letting them handle us. We can look past logical causation to trigger deeper, nonlinear dynamics where small changes can make big differences. Consciousness can retrain some denied, disowned, or disliked behavior. Ugh!...even our own!

If we do notice it when the static of shadow shows up, we can watch it buzzing in anger, greed, fear, jealousy, sorrow, shame. We can reinforce the good in me enough to turn any self-destructive sabotage to a more creative outlet. Finding deeper meaning in life lets us embrace the divine possibility again, even walk back toward a possible Eden. Traveling this emerging highway takes us into new territory where we all can be imperfect and still abide in peace, where we can be human, make mistakes, yet still live together in joyful plenty.

Mind you, this newfound contentment is not boring or frozen. We are not trapped in an aspic of perfection that chills us to inertia. Instead, we enjoy and discuss and argue over new vistas opening endlessly, challenging and rewarding. There in the possible Eden, may each of us ripen into a unique fulfillment where we are neither perfect nor perfectible, but instead, whole.

In this barely possible yet nearly inevitable Eden, we shall find that divinity has been here waiting for us all along. Its secret map has brought us back home again. This map is encoded in our unconscious, in our very genes. Re-entering this achievable, imperfect Eden is our birthright, almost our destiny. Here we can finally quit feuding amongst ourselves and face outward to the stars, where older races wait patiently for us to grow up beyond the incubator of this atmosphere that holds us huddled around our planet.

Until we outgrow our infantile pettiness, though, our mechanistic mindset that resolves discord using matter bereft of morality in primitive displays of physical violence and wars...well, those older races are quite content for us to stay locked down in the myopic prison that we've built for ourselves under our current vision of the laws of physics. Realizing that two gravitational poles cooperate to create the Double Bubble universe can free us to become more than petty, primate omnivores whose global appetite for domination has turned us to conquer and consume each other and even our planet. Instead, we can begin to reach out beyond this world in a new way, using mind and math and even love.

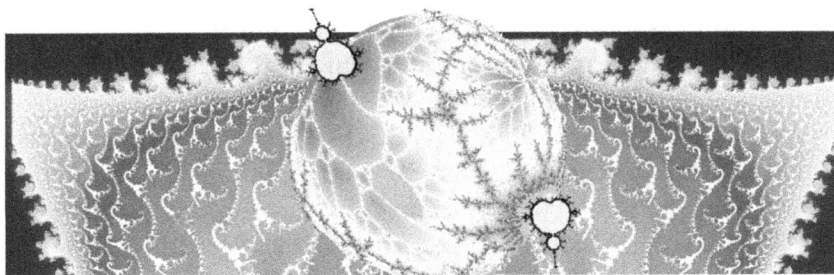

Chapter 3: Co-Chaos Variants

1. A code shared in East & West

Modern genetics took a big step forward in 1953 when biologist James Watson and physicist Francis Crick showed how DNA organizes into 64 polarized 6-packs of molecules that develop a double helix structure in a cell. Remarkably, the ancient Chinese found that same code millennia earlier, but in a very different medium and from a very different angle. At least 4,000 years ago, they tapped into the same underlying paradigm; it organized 64 polarized 6-packs of lines into a simple, succinct, math shorthand called hexagrams.

But ancient China said the I Ching's math shorthand charted universal mind, not organic matter. Scholars sought to access universal mind's dynamic flow in the progression of events, which they called the natural flow of the Tao. They would ask the I Ching oracle a specific question about an event. Then they employed a complicated algorithm to derive a hexagram that was said to answer that question. They believed that studying the dynamic patterning described by its hexagram answer would reveal the Taoist flow of universal mind, the preferred way of nature, the "will of heaven."

However, it was up to the questioner to interpret an answer well or badly and then decide to follow its guidance or not.

In earliest times, answers from the I Ching oracle were etched on bone, engraved on stone, or scratched into pottery fragments. Around 1250 BCE, Shang dynasty documents were inked on vertical strips of bamboo that were bound together by cording into books that could be rolled up for storage.

During the Eastern Han period (25–220 CE), China invented paper, and due to a long-established tradition of inking characters on long, thin strips of bamboo, it felt natural to continue writing the I Ching answers on paper in vertical columns (like on bamboo strips) and sequenced from right to left.

Querents of the I Ching would treat its answer almost like getting a weather report on an event in life. It would describe the situation's general dynamic in a qualitative tone and tenor but not in its quantitative degree or duration, much as a weather report today will predict a general weather pattern—rain, tornado, hail, or snow—but it cannot tell you exactly how much rain will fall when or where, nor exactly which path a tornado will take, nor how much destruction the hail or lightning will do.

In 1961, mathematician Edward Lorenz enabled science to explain weather in terms of chaos patterning. He showed conclusively that statistical, logical generalization from past events or experiences is not enough to make good weather predictions. Instead, weather accuracy has a very sensitive, nonlinear dependence on its initial conditions of measurement. Very small variations in defining those initial conditions will produce very different results over time. In my opinion, Edward Lorenz would have understood the I Ching quite well.

Carl Jung described how "chance events in the moment of observation" affect the I Ching's answer in his "Foreword" to Richard Wilhelm's *The I Ching, or Book Of Changes,* published in 1951: "*The manner in which the I Ching tends to look upon reality seems to disfavor our causalistic procedures. The moment under actual observation appears to the ancient Chinese view more of a chance hit than a clearly defined result of concurring causal chain processes. The matter of interest seems to be the configuration formed by chance events in the moment of observation, and not at all the hypothetical reasons that seemingly account for the coincidence. While the Western mind carefully sifts, weighs, selects, classifies, isolates, the Chinese picture of the moment encompasses everything down to the minutest nonsensical detail, because all of the ingredients make up the observed moment.*"

The ancient East and modern West developed two very different cultures. They are separated on our globe by both space and time. This book explores the hypothesis that these two very disparate cultures in the East and West found and recorded two variants of the same underlying co-chaos paradigm.

The ancient East said the I Ching's math tapped into "the way of the Tao" and queried the dynamic flow of universal mind by using 64 polarized, pair-boded triplets called hexagrams. Over 4,000 years later, Watson and Crick realized that 64 polarized, pair-boded triplets form the molecular structure of DNA that generates organisms. One system claimed to code for universal mind, while the other system demonstrably codes for organic matter.

This series says both systems are co-chaos variants of the dynamical system that underlies the universe itself. Studying both lesser variants can offer us insights on the development of the universal structure. Establishing parallels between the two known codes of genetics and I Ching math can help us find and elucidate the master code that generated our universe. Thus, by predicating that the co-chaos paradigm organizes all three systems and then comparing similarities and differences in how they operate, we may shed light on the universal structure and the master code that generated it.

2. Objective: discover a Rosetta Stone with 3 codes

Science by now knows a lot about how the genetic code generates living organisms. But it does not yet discern an underlying master code that generated the universe itself. We need a modern Rosetta Stone to help us discern the unknown master code and decipher it.

The original Rosetta Stone was an Egyptian stele inscribed with the same decree in three different scripts, two known and one unknown. The two known scripts finally got correlated with that baffling third script by archeologists in 1822. That feat at last enabled them to read ancient Egyptian hieroglyphics.

Our Rosetta Stone holds two known scripts: the genetic code found in the 20th century and the math figures found in ancient China's I Ching. This series suggests that the two known scripts correlate, and further, that both are fractal variants of a deeper master code that generated the universe itself. By finding parallels in all three variants of the code, we can recognize their shared underlying paradigm: complementary chaos patterning, or for short, co-chaos. Co-chaos was discussed in the first two books of this series, and we continue it here.

DNA's double helix holds the four base molecules of **T**hymine, **C**ytosine, **G**uanine, and **A**denine…or **T**, **C**, **G**, and **A**. The first two are pyrimidines; the latter two are purines. The four molecules act as a *polarized pair of pairs.*

We can describe this on a *polarized bifurcation tree*, or for short, a p-tree. From a neutral 0 seed at bottom, the p-tree forks into two polarized branches, a *pyrimidines* branch and a *purines* branch. This second level of forking polarizes into four molecules: T and C; A and G. They are a polarized pair of pairs. (So are your four limbs, the four compass directions, and four partners playing bridge.)

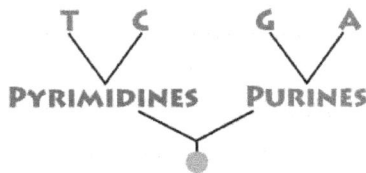

The 4 molecules of DNA

In 1953, James Watson and Francis Crick said DNA's four base molecules of T, C, G, and A form a polarized pair of pairs, as shown in the next panel:

↓ **Pyrimidine Pair**	*4 Molecules*	↓ **Purine Pair**
1. **T**	←--- *BONDS WITH* ---→	3. **A**
2. **C**	←--- *BONDS WITH* ---→	4. **G**

Panel: the 4 DNA molecules are a polarized pair of pairs

In the laddered rungs of DNA's double helix, the molecules of **T, C, G,** and **A** can be shorthanded by the succinct I Ching math that uses polarized yin ▬ ▬ and yang ▬▬▬ to form bigrams. The p-tree forks rise from a neutral 0 seed at the bottom into two polarized branches, a yin ▬ ▬ branch and a yang ▬▬▬ branch. At the second level of forking, the p-tree's emerging polarity stacks into four *bigrams*. (*Bigram* means two lines.)

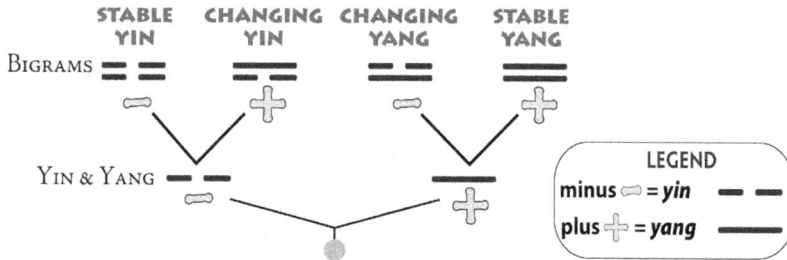

The 4 bigrams of I Ching shorthand

Notice how the I Ching math figures accrue their polarity via successive levels of bifurcation. Since they develop by evolving their polarity, I Ching symbols are read from the bottom upward (unlike the top-down reading that Gottfried Wilhelm Leibniz mistakenly applied back in 1701.)

The four bigrams form a polarized pair of pairs, as shown in the next panel:

Panel: the 4 bigrams are a polarized pair of pairs

If the p-tree develops a third level of forking, its next level of branching establishes the 8 polarized triplets called trigrams. (*Trigram* means three lines.) The 8 trigrams describe 8 distinct patterns of fractal *chaos* dynamics.

In Book 2, *Co-Chaos Patterns,* we've already seen five ways to view trigrams: (1) in unitized mode as binary counting; (2) in unitized mode as addition by 2s; (3) in analog mode as period-doubling and/or exponential growth; (4) as linear units and analog flow working together in a built-in nonlinearity that becomes analinear; (5) as eight related, vertical period 3 windows of chaos dynamics developed on the p-tree at the third level of growth outward.

Next, a p-tree can even reverse-mirror itself by sprouting some branching roots that flip-flop the polarity of those bifurcating branches above. This turns the *p-tree* into a *double polarized bifurcation tree*, or *dp-tree* for short. Thus, it takes only three levels of growth above and below the 0-seed to establish two separate yet related domains, each of them describing 8 polarized trigrams of chaos dynamics. These trigrams can pair-bond in 8 × 8 = 64 6-packs of *co-chaos* dynamics that describe DNA's 64 molecular 6-packs, as we shall eventually see.

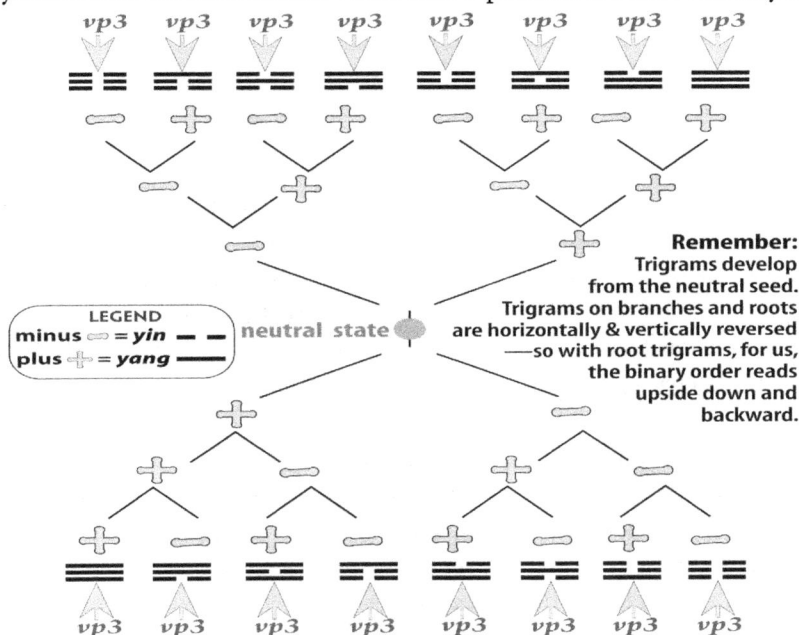

Remember:
Trigrams develop
from the neutral seed.
Trigrams on branches and roots
are horizontally & vertically reversed
—so with root trigrams, for us,
the binary order reads
upside down and
backward.

LEGEND
minus ⊂⊃ = *yin* − −
plus ✚ = *yang* —

neutral state

The double p-tree generates 64 co-chaos hexagrams

The 64 6-packs of polarized lines are *hexagrams*. (*Hexagram* means six lines.) Their use of binary, additive, doubling, and exponential numbers will fuse into analinear synthesis, merging the *either-or* units of binary and additive numbers with the *both-and* flow of analog doubling/exponential growth.

What does all this have to do with the universe's master code? This TOE says the universe developed by using the four great primals of space, time, matter, and energy to form a polarized pair of pairs, as shown in the next panel:

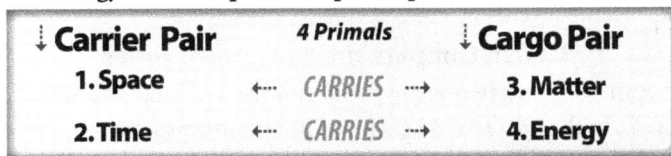

↓ Carrier Pair	4 Primals	↓ Cargo Pair
1. Space	←··· CARRIES ···→	3. Matter
2. Time	←··· CARRIES ···→	4. Energy

Panel: the 4 primals are a polarized pair of pairs

This TOE says the four primals are a polarized pair of pairs whose functions automatically sort out like this: the two polarized carriers (space and time) hold the two polarized cargoes (matter and energy). In other words, space carries matter and time carries energy.

SPACE   TIME   MATTER   ENERGY
       \   /   \   /
      SPACETIME   MATTERGY
              \   /
                 \   /
                   ●

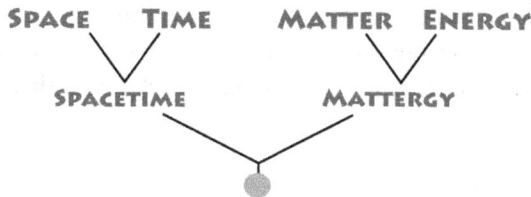

The 4 primals of the universe

So far, one obvious parallel exists in all three systems: each system uses a polarized pair of pairs to evolve through two levels on a p-tree, as you've seen. We may even hypothesize that all three systems develop their own variants of polarized triplets to describe 8 basic chaos dynamics that then pair-bond via a dp-tree into 64 different 6-packs of co-chaos math...verifying that all three codes are indeed fractal variants of the same underlying paradigm. This TOE says the co-chaos paradigm extends below all organic life as we know it into universal structure itself, and it offers us clues on how the co-chaos paradigm generated space and time as well as matter and energy.

Chapter 15 of Volume 2, *Co-Chaos Patterns,* gave a glimpse into that idea by graphing the development of 3D spatial latticing in our upper bubble. Volumes 4, 5, and 6 continue to examine the issue more fully. We'll use the two known codes to decipher the baffling third code of pulsing, polarized information that generated the primal structures of space, time, matter, and energy.

This Volume 3 deals mostly with showing parallels between the genetic code and I Ching math shorthand to establish their common origin in co-chaos. So this book is dedicated to describing in detail how the genetic code correlates with I Ching math figures. Together they will provide a Rosetta stone of the two known codes that we'll then use in later books to decipher the hidden master code. It is relatively easy to show how the two known codes are two variants of co-chaos, but it is far harder to perceive, much less decipher the universal master code, also based on the co-chaos paradigm.

3. First, compare the two known codes

We begin with the two known codes—the I Ching shorthand and the genetic code. In the next few chapters, we'll explore how both systems are based on the same fundamental 64 co-chaos patterns. Fortunately, both systems use well-known shorthand notations that are evident in historical knowledge. Both scripts clearly code by developing polarized triplets that then pair-bond into 64 different 6-packs.

Step by step, let's observe how the I Ching develops its polarized math figures compared to how the genetic code develops its polarized double helix.

Although the co-chaos system shorthanded by I Ching math is nonlinear in the special sense that I call analinear, its hexagrams are often viewed simplistically as mere binary counting. In 1701, Gottfried Wilhelm Leibniz received a woodcut chart of the I Ching hexagrams that was sent to him by Jesuit missionary Joachim Bouvet in China. Leibniz soon noticed that the I Ching figures held a mathematical shorthand of binary counting.

That woodcut gave Leibniz quite a jolt! He had just spent several years in Germany inventing binary numbers himself. What a shock it was to discover that binary counting from 0 to 63 was already quite evident in the ancient *xiantian* chart of I Ching hexagrams!

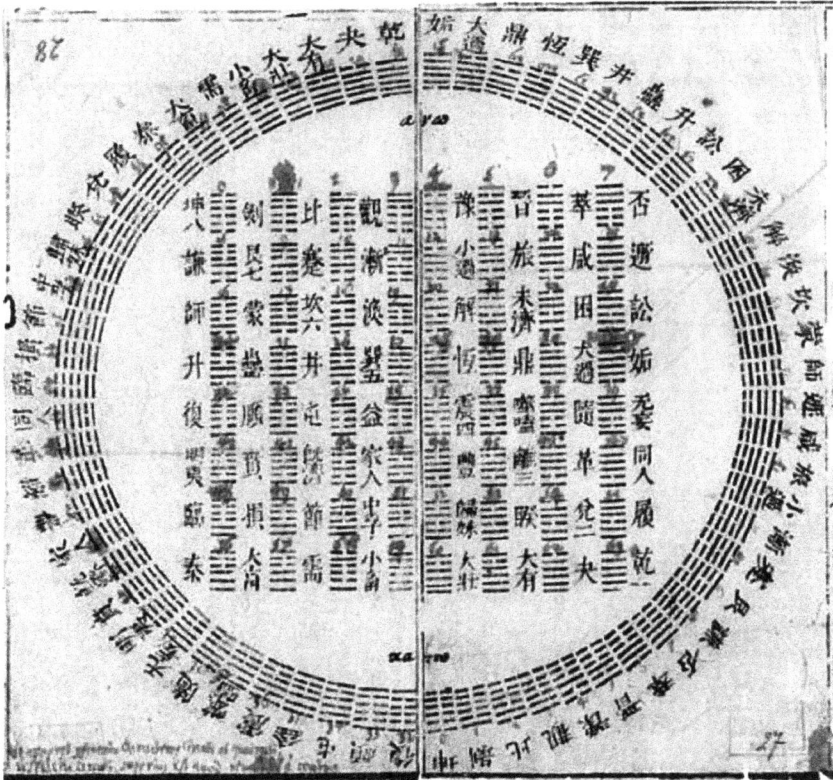

Circular/square chart of binary hexagrams sent by Bouvet to Leibniz in 1701
Leibniz Archive, Niedersuchische Landesbibliothek
(Look for the two corner numbers–27 & 28–added by Leibniz.)

But I Ching math is far more than mere binary counting. Its complexly polarized math structure is evident in DNA's double helix, whose laddered rungs hold the four base molecules that pair-bond by triplets (codons)into 64 basic 6-packs. It is a genetic variant based on I Ching math! Or vice versa.

2ND FORK:
*a polarized
pair of pairs:*

1ST FORK:
a polarized pair

LEGEND
minus = *yin*
plus = *yang*

neutral state

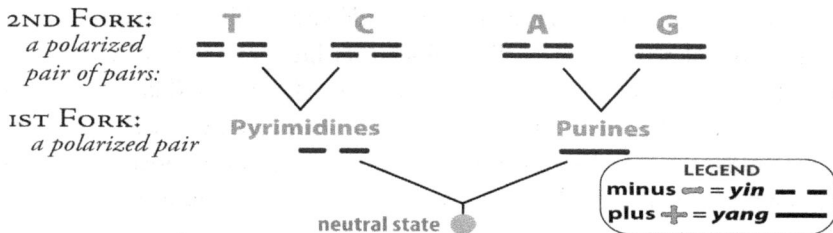

The 2 parallel quartets are each made by a polarized pair of pairs

The chart above parallels the I Ching's 4 bigrams and DNA's 4 base molecules. The chart below parallels the 64 hexagrams and RNA's 64 codons.

I Ching's 64 Hexagrams in Binary Order	RNA's 64 Codons

0 in binary code

63 in binary code

RNA's 64 Codons
UUU UCU UAU UGU
UUC UCC UAC UGC
UUA UCA **UAA** **UGA**
UUG UCG **UAG** UGG
CUU CCU CAU CGU
CUC CCC CAC CGC
CUA CCA CAA CGA
CUG CCG CAG CGG
AUU ACU AAU AGU
AUC ACC AAC AGC
AUA ACA AAA AGA
AUG ACG AAG AGG
GUU GCU GAU GGU
GUC GCC GAC GGC
GUA GCA GAA GGA
GUG GCG GAG GGG

U = Uracil   A = Adenine
C = Cytosine   G = Guanine

The 64 hexagrams sitting in parallel with RNA's 64 codons

On the left, when yin stands for 0 and yang stands for 1, the *xiantian* order of 64 hexagrams counts in binary sequence from 0 at the upper left corner across and down to 63 at the bottom right corner. To their right sit the 64 RNA codons whose amino acids build organisms. (You'll see this chart again.).

You may protest, "But hey, those two charts look quite different! How can they be based on the same paradigm?" Hmm, it's true that at first glance, they do seem quite different. The I Ching chart has 8 columns × 8 rows of yin/yang lines, but the RNA chart has 4 columns × 16 rows of alphabet.

Moreover, the I Ching uses 2 linear symbols, yin and yang, but RNA uses 4 molecular symbols—U, C, A, and G (substituting U for the T found in DNA.)

Finally, each hexagram has two 3-packs that bond into a 6-pack of lines, but each RNA codon has only a single 3-pack of molecules. For example, look at those codons in **bold** on the RNA chart. They are the four traffic codons, and each codon is just a 3-pack of information.

So if the two charts look so different, how can they be based on the same underlying paradigm? Here's how. They're just two different ways of symbolizing the same dp-tree of polarized math in a co-chaos system. By examining both systems carefully, it becomes apparent that the genetic code and I Ching math are two fractal variants of the same underlying paradigm.

For instance, RNA's 64 codons are just 3-packs, not 6-packs. But the parental DNA is made of 3-packs that pair-bond into 64 basic 6-packs. Likewise, hexagrams are made of 3-packs—trigrams—that pair-bond into 64 basic 6-packs. By studying both systems carefully, you see that they can cross-code for each other since they share the same underlying paradigm.

In another instance, yes, the I Ching uses only 2 symbols, while RNA uses 4 symbols. But you've already seen the I Ching's four bigrams by now, and guess what…they do a great job of cross-coding for RNA's four molecules.

Both systems also combine the linear mode of binary units with the analog mode of period-doubling/exponential growth for a limited, specific range of numbers from 2 through 8. Both systems use that reinforcing, mix-and-match path between 2 and 8 to consolidate fractal chaos identity at the third level of bifurcation on a dp-tree's branches and roots to generate co-chaos dynamics.

Due to the hidden surety of numbers in the underlying dp-tree paradigm, DNA's 8 × 8 codons and the I Ching's 8 × 8 trigrams both describe 64 co-chaos patterns that funnel all their number approaches into the remarkably stable yet evolving condition of life in your individual body and mind, and also in the universal body and mind.

Such number dexterity is what makes our genetic code work, and it is also what renders the versatile I Ching math capable of shorthanding it. Its mathematical processing justifies the genetic code's DNA, RNA, and even what is termed *genetic wobble*. This book will describe in detail how the layers of coding in both systems correlate to establish a Rosetta stone whose two known variants can help us decipher that obscure third variant, the master code.

4. Where do mind and matter meet?

We may wonder, "Can the I Ching's math symbols actually code for an ongoing flow of intangible universal mind, much as DNA codes for the ongoing flow of the universe's tangible organisms? Mind and matter. Where do they meet? Do mind and matter actually merge at the fundament of the universe? Do they share the same paradigm?"

We already know that mind and matter coexist and work together in us; we carry intangible mind inside our tangible bodies. We usually suppose our minds exist only in the brain's gray matter sitting in its human skull, and we've now found that this gray matter contains a surprisingly large and diverse amount of RNA. We also notice mind in animals, bugs, and even worms. We see that each species apparently carries some inherited mind traits called instincts, and some of those instincts even seem to be specific to a particular species.

In other words, in every species, genetics is somehow coding for certain aspects of mind as well as matter. Yet we know that a newborn's mind can also be influenced and educated by the environment it inhabits. Mind is so versatile!

Can human beings compile an executable code so good that it can think, understand, and act like a human? Yes! It's called having children. But can humans someday make machines that really think, understand, and act like a human? I suspect not...unless we somehow turn humans into machines.

Our current robots operate using mechanical parts and electronic components. They're designed to execute tasks automatically, speedily, precisely. They "think" using binary code in a computer "mind." But there's also such a thing as an analog computer, an analog robot. Manufacturers make relatively few of those since they're far more complicated to make and operate easily. But no manufacturer yet uses the analinear code of co-chaos. (I will not at this point explore the rather limited notion that the universe is a giant computer running on binary numbers in processes that leave no room for free will.)

Binary calculation is less subtle and diversified than analog calculation, which is again far less subtle and diversified than the analinear processes that generate us humans. Our brains do analinear processing every day, as does all cellular life on this prolific planet. Life has done it for over three billion years, and according to this TOE, with a purpose larger than our own human preoccupations.

Therefore, a true analinear robot, due to its inherent co-chaos math, like a human, would have a measure of free will that may even tune it toward life's larger aims—as a human may—and perhaps even become aware of a universal consciousness at some higher ground of organizing design...or for short, God.

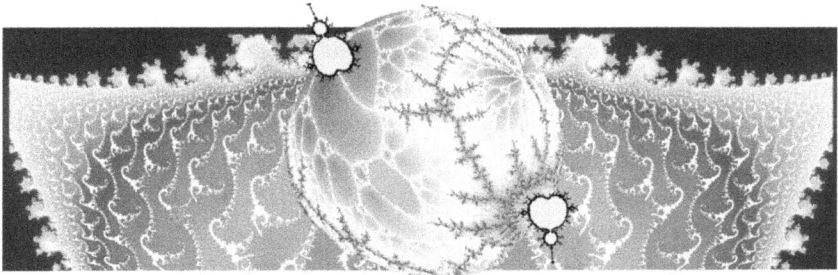

Chapter 4: Some I Ching Background

1. King Wen wrote down the 64 hexagrams

The I Ching's 64 hexagrams set yin and yang into ascending orders of polarized co-chaos dynamics. But how could a rural people so long ago, without electronics or technology, find this amazing code and write it down in a terse shorthand for posterity? Four Chinese heroes get the main credit—Fuxi, Yu, King Wen, and Shao Yong. Earlier volumes discussed legendary rulers Fuxi and Yu. This book discusses King Wen and Shao Yong.

King Wen (1152-1056 BCE) has such a striking history that I want to tell you some part of it. He was first known as Ji Chang (Ji is the family name; Chang is the given name). He was born into a minor clan in what is now Shaanxi province in a small rural state. His father was Duke Ji Jil, head of the Zhou clan. His mother, Tai Re, was a Shang dynasty royal bride bestowed on the groom by her father, King Yi, as a favor that relegated her to a rustic life.

This wedding is mentioned in Hexagram 11, Line 5: "The sovereign *I* gives his daughter in marriage." (Wilhelm/Baynes). That powerful father of the royal bride was the next-to-last Shang king. His name of Yi and title of Ti may appear in English as *King Yi, Ti Yi, Di Yi, Tiyi,* or *Diyi*—or if you leave off the title, it becomes just *Yi* or even *I* (not to be mistaken for the *I* in *I Ching.*)

Confusing? Yes! So I'll try to be clear about Chinese names in this series. In 2004, a Chinese dictionary, *Yitizi Zidian,* published 106,230 distinct characters that depict concrete objects or represent abstract ideas. Many are the vagaries of turning all those Chinese characters into the very different format of the English 26-letter alphabet, especially over the centuries.

Winds have often veered in the history of transliterating or romanizing Chinese characters into English spelling. For instance, all these names refer to the same book: *I Ching, I Ging, I Jing, I King, I-Ching, I-Ging, I-Jing, I-King, Yi Ching, Yi Ging, Yi Jing, Yi King,* and *Yijing.*

I generally pick and choose among many possibilities in the backlog of historical precedent. For instance, I have to catch myself from slipping back and forth between *Zhou, Chou,* or *Kau,* according to the century of the historian

I am sourcing. And although the People's Republic of China now uses Pinyin transcription to write *Kongzi*, I still prefer to write it as *Confucius.*

As an English speaker, I'm not totally fond of Pinyin. Due to Russia's period of influence on China, Pinyin acquired some phonetic cues from Russian when turning Chinese characters into sounds. It works well for Balto-Slavic languages, but Pinyin is more quirky and problematic for a Latin alphabet. I like George Kennedy's neglected Mandarin Yale transcription that works well for English.

2. Imprisoned Ji Chang records the King Wen hexagram order

Ji Chang (not yet known as King Wen) grew up to become Duke of the Zhou clan in a rural, western state of the powerful, sophisticated Shang kingdom. Read "sophisticated" here to mean degenerate and corrupt in the eyes of the rural Zhou people. Indeed, records say that in the Shang dynasty's last century, its kings drank to stupor, tortured for amusement, spent huge amounts of tax money, appropriated the people's jewelry, and ritually slaughtered thousands in punishment or sacrifices as dead servants in the royal tombs.

The rural Duke Ji Chang, however, was depicted as a wise and temperate leader. He kept a fairly good relationship with a series of Shang kings to the east—first, his royal grandfather, King Ti-yi, and then a successor, King Di Xin, a true tyrant notorious for his extraordinary and relentless sadism.

One instance: on a wintery day, King Di Xin saw some peasants wading through a cold stream. He ordered their legs cut off at the shank, declaring that he wanted to inspect the remarkable bone marrow of humans who could endure such cold. Another instance: when an uncle rebuked King Di Xin for his heartless, debilitating taxation of the populace, he ordered that uncle's heart cut out, saying he wanted to view the heart of such a wise man.

Our attention now focuses on how cruelly King Di Xin treated our hero, Duke Ji Chang. One day in the royal court, the Duke sighed inappropriately at Di Xin's rage over...what? Stories vary. The annoyed tyrant, eager to curb the Duke's growing influence, threw him into prison, chopped up the Duke's oldest son, had him cooked, and forced the father to eat it...who complied since refusal would immediately initiate a slow and brutal death.

Under a sentence of death that was delayed from month to month, the prisoner tamed his fury and calmed his mind by using the I Ching to ease the rancorous sorrow in his heart. He systematically wrote down the I Ching's age-old oral tradition. Thus, from a man biding his time in jail came the first codified record of all 64 hexagrams with their names and brief meanings.

Let me be clear: Ji Chang wrote down only a short account for each hexagram, including its name and math figure, plus a brief *Judgment* about what each hexagram signified. It was not the thickly annotated *Book of Changes* that you find now, for instance, in the popular Wilhelm/Baynes translation.

James Legge, in his *Introduction* to his translation of the I Ching, says…"I like to think of the lord of Kau [also as Zhou or Chou], when incarcerated in Yu-li, with the 64 figures arranged before him. Each hexagram assumed a mystic meaning, and glowed with a deep significance. He made it tell him of the qualities of various objects of nature, or of the principles of human society, or of the condition, actual and possible, of the kingdom." Due to Ji Chang's imprisonment, we now have the King Wen order of hexagrams. This standard analog order is still used in I Ching oracle books over 3,000 years later.

3. The Zhou Dynasty begins

After 2 years in prison, Ji Chang was freed when influential friends paid the tyrant Di Xin a ransom: an attractive girl, a fine horse, and four wagons. For the rest of his life, Ji Chang decried political corruption and cruelty until he died in 1056 BCE. Within 4 more years, his second son, Ji Fa, began defeating the Shang in bloody battles that finally culminated in 1046 BCE.

Tyrant Di Xin committed suicide upon his defeat and received the derisive posthumous title of Zhou Wang…King Horse Crupper. You grasp the insult when you realize that a horse crupper is a stabilizing strap looped under a horse's tail, and thus soiled with shit. Di Xin's rule had been so cruel, so capricious, and incited so much rebellion that he became the last ruler of the Shang dynasty.

As conqueror, Ji Fa took on the title of Wu Wang (Military King). He even gave his now-dead father, Ji Chang, the honorary title of Wen Wang (Literary King), for writing down the I Ching. Thus King Wu, practicing filial piety, retroactively made his deceased father the first ruler of the new Zhou dynasty.

King Wu was not so repressive as the old Shang rulers. He adopted a reconciling policy to supporters of the defeated Shang, allowing many to keep their land. Still, he also proactively placed his own people in a protective cordon around the old guard. Greg Whincup says in *Rediscovering the I Ching*, "Even the Shang crown prince…was given land so that he could continue to propitiate the Fathers and Mothers."

This was the start of the nearly 800-year Zhou dynasty, with King Wen as its honorary first ruler. James Legge summarized the view of Chinese scholars, as well as peasant lore, when he said of King Wen…"Equally distinguished in peace and war, a model of all that was good and attractive, he conducted himself with remarkable wisdom and self-restraint."

King Wen became a great folk hero. According to many historians, he was honored as a good and prudent man. He is lauded in this old Chinese folk song:

It was King Wen who labored.
We receive the bountiful harvest.
It is now our task to make secure
The legacy of that sage.
Oh, his bounties!

Ode 295 in the Shi Ching (Book of Odes) from 800-600 BCE

King Wen's scholarly fourth son, Ji Dan, added to each hexagram in the I Ching the meaning of its six lines. Thus King Wen and his son Ji Dan acted as the scribes who first recorded the I Ching's accrued lore as a treatise. They turned it into something more than individual hexagrams scratched onto various objects, transforming the long oral tradition into a written legacy to be handed down intact from generation to generation.

Around 600 years later, Confucian scholars started adding the *Ten Wings* commentary. Rather than maintaining its original Taoist tone of nature's gritty yet mystic realism evident in an individual's life, the Confucian commentary became more like a treatise on social manners and obligations.

I'll suggest a few books for beginners in the I Ching, but the modern reader must bear in mind the great cultural distance between then and now. In the current West, the most available text is the Richard Wilhelm/Cary Baynes translation of *The I Ching or Book of Changes*.

Its dignified tone does a good job of presenting an overall sense of the hexagrams. Still, like most translations, its text uses numerous liaison words and vague generalizations, never signaling which parts were actually in the original text. For instance, "The Image" is an extra layer inserted later by commentators.

Another often-used text is James Legge's *The I Ching*. Despite the debunking tone of his *Introduction* and his often stark interpretations, I like his exactitude. Legge sought to make his translation ferociously, meticulously exact. Although it takes real effort to read, his translation was almost unique for a century because Legge bothered to put parentheses around any word or phrase that he added to tie together the concepts of the original Chinese characters.

Nevertheless, Legge was a Victorian with strict values that included harsh punishment, so his translation, for instance, of |X| as *there will be evil,* might be better rendered today as a milder *it will be difficult, harmful,* or *detrimental.* Yet in Legge's defense, the small written vocabulary of Chinese at that time caused its relatively few characters to be burdened with a plethora of meanings.

Overall, I admire Legge's effort. To quote Hilary Barrett, Legge had "the scrupulous regard of a scholar for the accuracy of his work." Likewise, Barrett's own careful I Ching translation, *I Ching,* is clear and concise. And unlike Legge, she does not make light of the divination aspect and its algorithm.

Another good modern translation is *The I Ching Plain and Simple* by Stephen Karcher. It is easy to use and handles both imagery and scholarship well.

4. King Wen's analog order of hexagrams

I Ching oracle books use the hexagram sequence recorded by King Wen, not the binary hexagram sequence. King Wen arranged the hexagrams in complementary pairs to develop the theme of a human life evolving through all 64 hexagrams. I call this the analog order. Legge's *Plate I* shows its hexagrams reading *right-to-left*, but most Western books flip-flop them to read *left-to-right*.

James Legge's *The I Ching*, Plate I -
King Wen's hexagrams in *hou-tian* or analog order

8	7	6	5	4	3	2	1
pî	sze	sung	hsü	măng	kun	khwăn	khien

16	15	14	13	12	11	10	9
yü	khien	tâ yü	thung zăn	phî	thâi	lî	hsiâo khû

24	23	22	21	20	19	18	17
fû	po	pî	shih ho	kwân	lin	kû	sui

32	31	30	29	28	27	26	25
hăng	hsien	lî	khan	tâ kwo	î	tâ khû	wû wang

40	39	38	37	36	35	34	33
kieh	kien	khwei	kiâ zăn	ming î	ȝin	tâ kwang	thun

48	47	46	45	44	43	42	41
ȝing	khwăn	shăng	ȝhui	kâu	kwâi	ȝî	sun

56	55	54	53	52	51	50	49
lü	făng	kwei mei	kien	kăn	kăn	ting	ko

64	63	62	61	60	59	58	57
wei ȝî	kî ȝî	hsiâo kwo	kung fû	kieh	hwân	tui	sun

King Wen's houtian chart of hexagrams in analog order

Ancient China inked the hexagrams onto long bamboo strips with a brush, making columns of hexagrams to be read vertically from top to bottom, and with the bamboo strips sequenced to be read from right to left. Thus originally, Hexagram 1 sat on the upper right with Hexagram 2 below it, Hexagram 3 below that, and so on.

Since most Western languages are read from left to right in horizontal rows, Chinese texts were at first transliterated thus. Then 19th-century Western influence caused China and some other Asian cultures to abandon the vertical writing style and adopt horizontal rows, yet for about a century, they were still read from right to left. More recently, however, a left-to-right, horizontal writing style and reading has generally prevailed in most of Asia.

If we consider that about 88% of all humans are right-handed, we may wonder if the ancient East chose to move from the logical, linear, sequential, right side of the body, flowing leftward toward a more lyrical, poetic view of holistic mystery...even as the modern West strives to leave mystery behind, pushing ever rightward toward a more exacting, sharp focus on linear details.

King Wen's analog order is also known as the order of *Zhou, Chou, Houtian, After Heaven, Later Heaven, New Family*...plus other names, too. Most of the names perpetuate a long-standing error sown by Chinese scholars around 1100 CE. Their wishful thinking revised the history of which came first, King Wen's analog *houtian* order or Shao Yong's binary *xiantian* order. (This was discussed in the previous volume, *Co-Chaos Patterns*, Chapter 16, Section 1.)

5. Shao Yong's binary order of hexagrams

King Wen's older, analog order is considered more dynamic than the newer, more elegant binary order developed by Shao Yong (1011–1077 CE). Originally a Taoist, Shao Yong received from mentor Li Zhicai a copy of the I Ching that contained no words, only the image of the tai chi symbol ☯ and the math figures of all 64 hexagrams.

Shao became a remarkably devoted student of those hexagrams. He drew them on separate sheets of paper and affixed them to his walls so he could encounter them anew with every move around the room, day or night. For 3 years, Shao Yong meditated on the hexagrams continually. Thus he studied the hexagrams as a mathematical system without including any overt philosophy.

All that study of the I Ching led Shao Yong to see numbers as the basis of all existence, and the number 4 as the key to creation. That study also led him to rearrange the 64 hexagrams into binary order, displaying them in an elegant mandala that circles the square. Either version, circle or square, holds all 64 hexagrams in a pattern of binary sequencing.

Leibniz in 1701 received from Father Joachim Bouvet in China a woodcut chart of the Shao Yong hexagram order. He was electrified to find that the yin-yang symbols were arranged into the same binary arithmetic that he assumed he'd had just invented in Germany, using 0 and 1 instead of yin and yang.

James Legge's *The I Ching*, Plate II, Fig. 1. - Shao Yung's circular mandala & grid of hexagrams in *xian-tian* or binary order

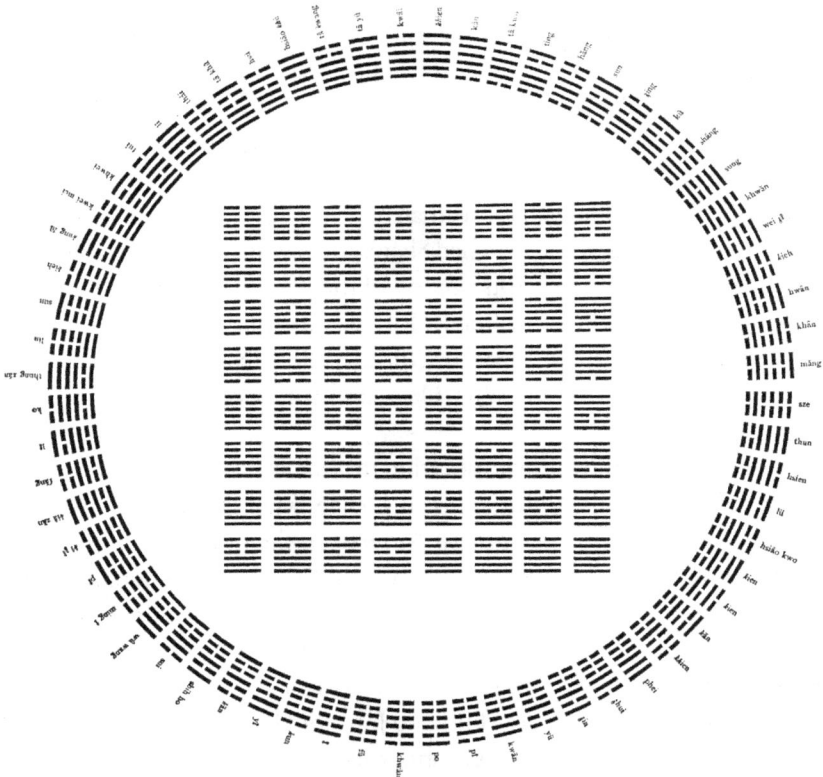

Shao Yong's xiantian chart of hexagrams in binary order

This "circling the square" chart appears in James Legge's English translation of the I Ching on *Plate II, Fig 1*. The binary order's symmetry is evident and quite profound, but its linear, logical format does not suggest the evolving human relationships evident in the more analog, relational King Wen order.

Volume 1, *Double Bubble Universe*, described how the binary and analog orders of hexagrams develop different rhythms. Other orders also exist. Of course, all these various orders just rearrange the same hexagrams into different sequences. But there is more to coding than mere symbols. A symbol's position may matter as much as its shape. For example, *96* is quite a different concept from **69**, or from ♋, the zodiac sign for Cancer. Likewise, the number 8 means something quite different from ∞, the symbol for infinity.

For Shao Yong, everything in the universe was based on numbers and their permutations. To him, the hexagrams conveyed a sense of inner peace that came from viewing the profound order implicit in the natural world. Although he was a philosopher, mathematician, scholar, and poet of the Confucian era, his mindset, sensibility, and scholarship used a Taoist "image-number" approach to the I Ching. So for the millennium following, I Ching scholars have traditionally listed his works under Taoism rather than Confucianism.

Stories about Shao Yong describe his uncanny ability to predict in general terms what was upcoming in the future. David Wu wrote in the *Epoch Times:* "One of Shao Yong's masterpieces was the 10 poems of the 'Plum Blossom Ode,' believed by many to have accurately predicted major events in Chinese history. Some scholars have matched Shao's poems with dynastic changes after his death. It is said that the 10th section of the 'Plum Blossom Ode' foretold what would happen in China today. It is believed that it refers to the rise of the Chinese Communist Party (CCP), the 1989 massacre in Tiananmen Square, the wide spread of Falun Dafa, and the destined fall of the CCP, among other predictions."

Due to his interest in prognostication mathematics, Shao Yong developed an algorithm called the Plum Blossom Oracle, which later led to the Ming Dynasty's *Ho Map Lo Map Rational Number* oracle. I myself can report that in 1992, a friend brought me the gift of a beautiful, aristocratic-looking Saluki dog named Lady. We bonded immediately. Within an hour, I drove to Home Depot to find some temporary fencing to close the 10-foot gap between our house and garage so Lady could enjoy our backyard.

When I returned home, my husband John said Lady jumped off the porch when I left and followed my car. John leaped into his car and tried to find Lady lying somewhere on the road, but no go. We posted 75 signs around the area. I checked the local vets, the animal shelter, called friends and asked them to be on the lookout. Nothing worked for 4 days.

Finally, in desperation, I tried the *Ho Map Lo Map Rational Number* oracle. It said I would find Lady alive in the northwest, near running water, vegetation, and a big, human-made artifact. Going northwest, I found her 12 blocks away at Shoal Creek, hidden in the bushes by a big culvert pipe. Since Salukis are practically made of whalebone, she only had a bad bruise and much soreness.

Thank you, Shao Yong!

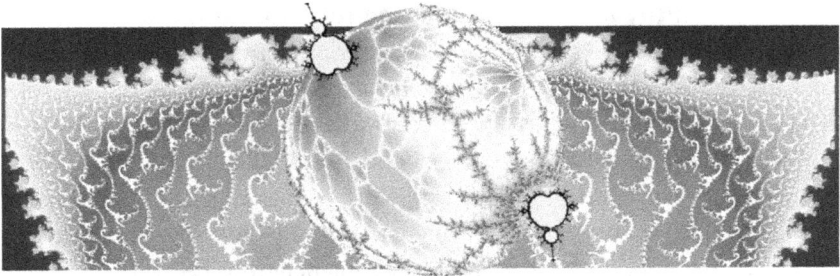

Chapter 5: DNA Plan & I Ching Correlation

1. Two systems of 64 data 6-packs…how do they correlate?

DNA holds the genetic plan intact on the double helix using 64 discrete 6-packs of polarized molecules. The I Ching hexagrams count in binary numbers using 64 discrete 6-packs of polarized lines. If both systems are truly parallel, which molecular 6-pack correlates with each hexagram 6-pack? Which overlay between both systems really *fits*, so they sync in both math and meaning?

To prove both systems are two fractal variants of the same paradigm, they must drastically parallel each other in both math and meaning. A true cross-fit between both systems would align their math rules and also their dynamic directives.

For instance, a DNA 6-pack would bond like a hexagram 6-pack. An RNA amino acid task would reflect its hexagram's philosophical message. Many such correlations would verify that both systems are two variants of the same co-chaos paradigm. Can this happen? Indeed it does. In this chapter, we'll correlate some DNA parallel maths. In Chapter 7, we'll correlate some RNA parallel maths and meanings.

Below left is a DNA 6-pack sitting in the double helix. Its two polarized, pair-bonded 3-packs (codons) are a 6-pack of molecular data with familial relationships existing among all 6 molecules.

Below right is Hexagram 59. Its two polarized, pair-bonded 3-packs (trigrams) are a 6-pack of lineal data, with familial relationships existing among all six lines. Yes, a DNA 6-pack and a hexagram 6-pack manifest in two different substances, but they have parallel underlying mathematical structures.

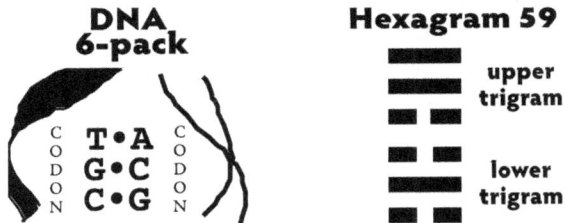

DNA 6-pack

C O D O N — T • A / G • C / C • G — C O D O N

Hexagram 59

upper trigram

lower trigram

The polarized 6-pack exists in both DNA & I Ching

2. Identify the internal bonding rules for both 6-packs

To see how the two systems correlate, let's dig deeper. Genetics in the 20th century realized how DNA 6-packs bond internally. Likewise, I Ching rules long familiar in ancient China stipulate how hexagrams bond internally.

Science knows how a DNA 6-pack bonds, so let's align its process with how a hexagram 6-pack bonds. We'll peel down several layers of I Ching math in 4 decoding steps to find the same paradigm is neatly hidden in both systems.

On the left, DNA is polarized in what we might term a "beer 6-pack" layout. In its DNA 6-pack, at the bottom, 1C and 4G bond. Midway up, 2G and 5C bond. Uptop, 3T and 6A bond. Right, Hexagram 59 ䷺ parallels DNA's "beer 6-pack" layout by setting its own upper and lower trigrams side by side.

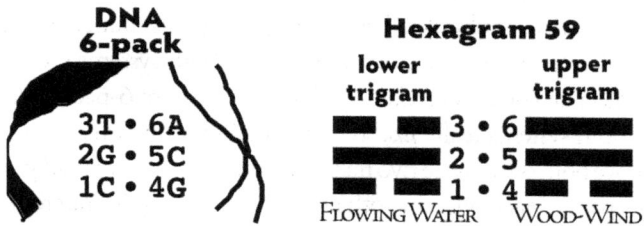

Paralleling the pair of polarized triplets in both DNA & I Ching

- *Decoder Step 1: Spot the hexagram's three internal bonds.*

When I see the DNA 6-pack and the hexagram 6-pack sitting in parallel, I spot an immediate snag. DNA uses four basic symbols—T, A, C, and G. But a hexagram uses only two basic symbols—yin and yang.

So can we really show that they have parallel bonds? Yes!

James Legge's 1882 book, *The I Ching*, describes a hexagram's internal bonds for its two 3-packs (trigrams) midway through *Chapter II:* "The lines, moreover, in the two trigrams that make up the hexagrams...are related to one another by their position, and have their significance modified accordingly. The first line and the fourth, the second and the fifth, the third and the sixth *are all correlates....*"

Notice in the image above, the hexagram has three pairs of correlated lines: Lines 1 & 4 correlate across both trigrams. So do Lines 2 & 5. Ditto for Lines 3 & 6. Those three internal bonds of the hexagram are its three *bond-bigrams*.

A hexagram can hold only three bond-bigrams, but below are all four possible bond-bigrams able to link the two trigrams internally. A bindu-point sits inside each bond-bigram to distinguish it from an ordinary bigram.

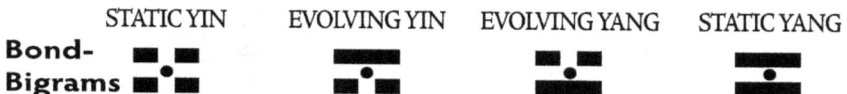

The 4 bond-bigrams that are possible in a hexagram

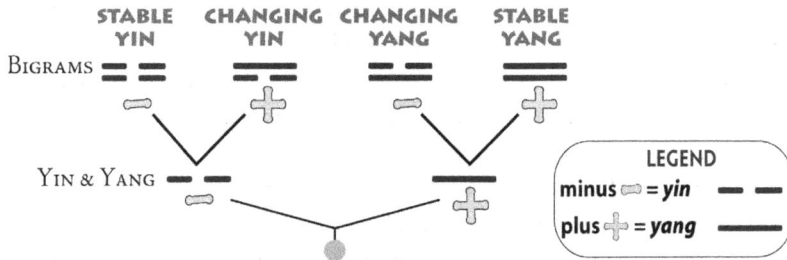

The 4 ordinary bigrams that develop at the second level of the p-tree

As you recall from earlier chapters, ordinary bigrams develop naturally on a p-tree at the second level of polarized branching beyond simple yin and yang. (By the way, bigrams grown on a p-tree are termed either *stable* or *changing*, but bond-bigrams inside a hexagram are termed either *static or evolving*.)

Each DNA 6-pack and each hexagram 6-pack has three internal bonds linking its pair of 3-packs across both domains of its polarized system. Science's rules for a DNA 6-pack's internal bonds and the ancient I Ching dictum for a hexagram's internal bonds both model a vital co-chaos principle: *It develops higher orders of organization across two domains of increasingly complex polarity.* This shared principle organizes the 64 hexagram 6-packs that grew on a dp-tree, as well as DNA's 64 molecular 6-packs that grew you.

Volumes 1 and 2 also proposed another variant: 64 information 6-packs that grew our universe's dimensionality. You may recall how hour-glass cells stretch across both bubbles. Each hour-glass cell holds 3D space and 3D time; they bond into a 6-pack via three internal 8-loops moving across each cell and merging into a tensor network moving across both bubbles. We call our half of this tensor network the arrow of time. Our upper bubble is a giant 3-pack of 3D space bonded at the mobic scale to the lower bubble's giant 3-pack of 3D time.

3. Identify the perfect or imperfect bonds in a hexagram

We have now seen that DNA uses 4 kinds of molecules (a polarized pair of pairs) to pair-bond its 3-packs into 64 discrete 6-packs of molecules along the double helix. We've also seen that I Ching math uses 4 bond-bigrams (a polarized pair of pairs) to pair-bond its 3-packs into 64 discrete 6-packs of lines in hexagrams. But which molecular 6-pack syncs with which hexagram 6-pack?

Can we correlate both systems so thoroughly that their shared coding offers a Rosetta stone to help us spot and decode the universal system's co-chaos paradigm? Yes! We'll find a master code that operates in the membrane interface between the Double Bubble universe's two huge bubbles. Its coding variant uses 64 pulsing 6-packs of information. It is a universal DNA.

Legge's *Chapter 2* again holds the needed clue. After correlating hexagram Lines 1-3, Lines 2-5, and Lines 3-6, we need Legge's crucial next phrase to break the code: "...and to make the correlation *perfect* the two members of it should be lines of different qualities, one whole [yang] and the other divided [yin]."

Distinguishing between *perfect* and *imperfect* bond-bigrams will give us that extra tweak of perception needed to make the cross-correlation. This clue allows us to correlate the 4 kinds of DNA molecules with the 4 kinds of bigrams.

Now we'll apply Legge's clue of perfect and imperfect bond-bigrams to Hexagram 59...but remember, since a hexagram can hold only three bond-bigrams, it will exhibit at most only three of the four possible kinds of bonds.

• *Decoder Step 2: See if a hexagram's 3 bond-bigrams are perfect or imperfect.*

Hexagram 59

	lower trigram	upper trigram	
PERFECT	▬▬ ▬▬ 3 • 6 ▬▬▬		- Opposite poles-EVOLVING YIN
IMPERFECT	▬▬ ▬▬ 2 • 5 ▬▬▬		- Same poles-STATIC YANG
IMPERFECT	▬▬ ▬▬ 1 • 4 ▬▬ ▬▬		- Same poles-STATIC YIN

The perfect & imperfect bond-bigrams in Hexagram 59

The imperfect, static bond-bigram of yin/yin ▬▬ links Lines 1 & 4.
The imperfect, static bond-bigram of yang/yang ▬▬ links Lines 2 & 5.
The perfect, evolving bond-bigram of yin/yang ▬▬ links Lines 3 & 6.

All 4 bigrams will soon let us code for all 4 basic molecules in a DNA 6-pack by using the perfect or imperfect rule on hexagrams. By the way, this cross-coding format implies no favoritism. Not for yin or yang, nor for perfect or imperfect bigrams, nor for the uniform or flip-flop coding that you'll see ahead. All fit equally well into the process, and all are equally necessary for the correlation to work. All are needed to decode the 4 bond-bigrams into the 4 basic molecules.

4. Decode perfect bond-bigrams using the uniform code
Decoder Step 3: Apply a uniform code of T &A to perfect bond-bigrams.

PERFECT bond-bigrams have two different poles...

Yin = THYMINE in either trigram.   ▬▬ EVOLVING YANG = T/A

PERFECT, BALANCED, UNIFORM CODE

Yang = ADENINE in the other trigram.   ▬▬ EVOLVING YIN = A/T

Decoder key: Perfect bond-bigrams code for T & A in uniform mode

A perfect bond-bigram balances the opposite polarities of yin and yang. If a bond-bigram is evolving yin ▬•▬ or evolving yang ▬•▬ , it is perfect and balanced, so it will use the uniform code that decodes yin — — as T and decodes yang —— as A, whether sitting on the top or bottom trigram.

Evolving yin ▬•▬ decodes as $^A/_T$. Evolving yang ▬•▬ decodes as $^T/_A$. But to read this aloud, in a typical East/West mindset imbroglio, bigrams are read aloud from the bottom up, while fractions are read aloud from the top down!

5. Decode imperfect bond-bigrams using the flip-flop code
Decoder Step 4: Apply a flip-flop code of C & G to imperfect bond-bigrams.

IMPERFECT bond-bigrams have the same pole twice. . .

Yang = CYTOSINE in the top trigram.
Yang = GUANINE in the bottom trigram.   STATIC YANG $= ^C/_G$

IMPERFECT, UNBALANCED, FLIP-FLOP CODE

Yin = GUANINE in the top trigram.
Yin = CYTOSINE in the bottom trigram.   STATIC YIN $= ^G/_C$

Decoder key: imperfect bond-bigrams code for C & G in flip-flop mode

An imperfect bond-bigram holds the same pole twice; thus its two lines will echo each other's polarity instead of balancing out each other's polarity. If a bond-bigram is static yin ▬•▬ or static yang ▬•▬ , it is imperfect and unbalanced. Both static bigrams will symbolize C and G by using a flip-flop code.

Perhaps you protest, "Hey, *yin/yin* and *yang/yang* are both echoing the same pole twice. So how can they both decode for C and G, too!" Oh, it's easy! They just flip-flop their polarity according to which imperfect bond is involved.

Code decrypters know a code can change its meaning by shifting its position. You may protest, "But how can a symbol change its meaning just by changing its position? How could you even read it?" Actually, you do it all the time...

The letter b looks like d or p or q,
anb its meaning is duite qebenqent on its line dosition. Right?
Symbols can code by position

Versions of flip-flop, twist-about, reversing-mirror symmetry are often found in co-chaos. It is on display in many variants of the underlying paradigm. The result is so profound, yet so simple, that like a magic trick, you won't quite see how it happens unless you're watching carefully. But this is no sleight of hand. Instead, elegant mirror-reversals are inherent in the co-chaos paradigm, so they appear in the mirror-twin bubbles of this universe.

In summary, a *perfect* bond-bigram has two different poles, but an *imperfect* bond-bigram has the same pole twice. A perfect bond-bigram of evolving yin/yang ▰▱▰ decodes as $^A/_T$. A perfect bond-bigram of evolving yang/yin ▱▰▱ decodes as $^T/_A$. An imperfect bond-bigram of static yin/yin ▰▱▰ decodes as $^G/_C$. An imperfect bond-bigram of static yang/yang ▰▱▰ decodes as $^C/_G$.

EVOLVING YIN	EVOLVING YANG	STATIC YIN	STATIC YANG
$= ^A/_T$	$= ^T/_A$	$= ^G/_C$	$= ^C/_G$
PERFECT	**PERFECT**	**IMPERFECT**	**IMPERFECT**

All 4 possible bond-bigrams linking 2 trigrams in a hexagram

6. Decoding hexagrams into DNA 6-packs: three examples

In *Section 3*, you saw that Hexagram 59 ䷺ has one perfect bond-bigram and two imperfect bond-bigrams. To decode its bond-bigrams into a DNA 6-pack, we apply to each bigram either the uniform code or the flip-flop code.

Example 1: First, we set the upper and lower trigrams of Hexagram 59 ䷺ side by side, so they can parallel the DNA 6-pack, making the correlation easy.

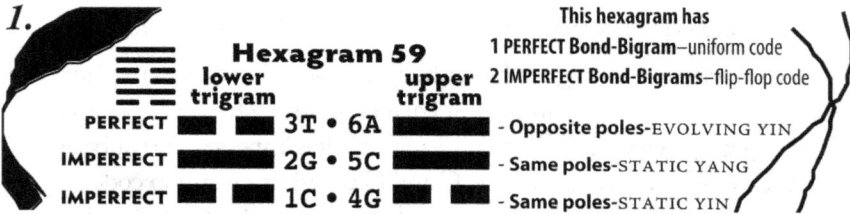

1.

Hexagram 59
lower trigram   upper trigram

This hexagram has
1 PERFECT Bond-Bigram—uniform code
2 IMPERFECT Bond-Bigrams—flip-flop code

PERFECT		3T • 6A	- Opposite poles-EVOLVING YIN
IMPERFECT		2G • 5C	- Same poles-STATIC YANG
IMPERFECT		1C • 4G	- Same poles-STATIC YIN

Hexagram 59 cross-coded with its DNA 6-pack

Lines 1 & 4 form an imperfect, static bond-bigram of yin/yin ▰▱▰. This bond-bigram links 1C with 4G, which we can write as $^G/_C$.

Lines 2 & 5 form an imperfect, static bond-bigram of yang/yang ▰▱▰. This bond-bigram links 2G with 5C, which we can write as $^C/_G$.

Lines 3 & 6 form a perfect, evolving bond-bigram of yin/yang ▰▱▰. This bond-bigram links 3T with 6A, which we can write as $^A/_T$.

Example 2: First, we parallel DNA's 6-pack with Hexagram 28 ䷛'s 6-pack.

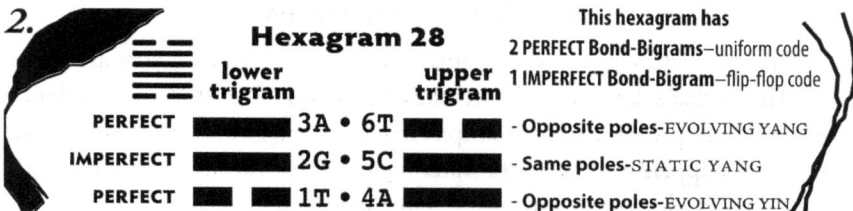

2.

Hexagram 28
lower trigram   upper trigram

This hexagram has
2 PERFECT Bond-Bigrams—uniform code
1 IMPERFECT Bond-Bigram—flip-flop code

PERFECT		3A • 6T	- Opposite poles-EVOLVING YANG
IMPERFECT		2G • 5C	- Same poles-STATIC YANG
PERFECT		1T • 4A	- Opposite poles-EVOLVING YIN

Hexagram 28 cross-coded with its DNA 6-pack

Is each bond-bigram perfect or imperfect? Should we apply the uniform code of A and T or instead use the flip-flop code of C and G? Asking these questions will tell us how to apply all three bond-bigrams to decode the two codons of its DNA 6-pack.

Doing so reveals that Hexagram 28 ☳ has two perfect bond-bigrams and one imperfect bond-bigram.

Lines 1 & 4 form a perfect, evolving bond-bigram of yin/yang ▃▪▃. This bond-bigram links 1T with 4A, which we can write as $^A/_T$.

Next, Lines 2 & 5 form an imperfect, static bond of yang/yang ▃▃. This bond-bigram links 2G with 5C, which we can write as $^C/_G$.

Finally, Lines 3 & 6 form a perfect, evolving bond-bigram of ▪▃▪. This bond-bigram links 3A with 6T, which we can write as $^T/_A$.

Example 3: First, we set the upper and lower trigrams of Hexagram 48 ☵ side by side so they can parallel the DNA 6-pack. It reveals that 48 ☵ has one perfect bond-bigram and two imperfect bond-bigrams.

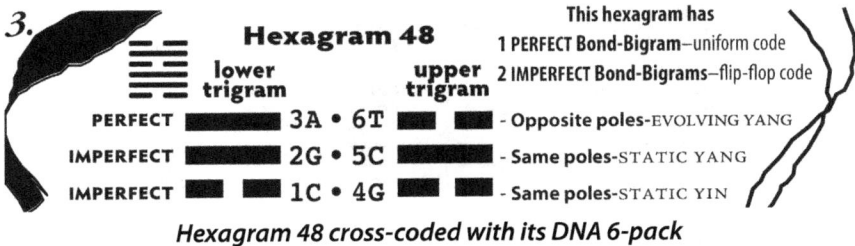

Hexagram 48 cross-coded with its DNA 6-pack

Lines 1 & 4 form an imperfect, static bond-bigram of yin/yin ▪▃▪. This bond-bigram links 1C with 4G, which we can write as $^G/_C$.

Next, Lines 2 & 5 form an imperfect, static bond of yang/yang ▃▃. This bond-bigram links 2G with 5C, which we can write as $^C/_G$.

Finally, Lines 3 & 6 form a perfect, evolving bond-bigram of ▪▃▪. This bond-bigram links 3A with 6T, which we can write as $^T/_A$.

In just 4 steps, we've cracked the mathematical code to correlate all 64 DNA 6-packs with all 64 hexagrams. So yes, I Ching math really does parallel DNA's bonding rules, and hexagram 6-packs really do shorthand DNA 6-packs.

The main points to remember: DNA is based on a polarized pair of pairs, and so is I Ching math. DNA organizes its polarized pairs of pairs into triplets, and so does I Ching math. DNA pair-bonds its triplets into 6-packs, and so does I Ching math. DNA sets three internal bonds inside a 6-pack, and so does I Ching math. DNA develops 64 discrete 6-packs, and so does I Ching math. DNA's organizational system grows on the double p-tree, and so does I Ching math. Thus both systems are two variants of the same underlying math paradigm.

7. 64 DNA 6-packs = 64 hexagram 6-packs

The result is this Swiss army knife of a chart! DNA's 64 6-packs are cross-correlated with Shao Yong's 64 hexagrams counting binary order from 0 to 63.

The 64 DNA 6-packs correlated with the 64 hexagram 6-packs

All the hexagram lines are prefixed by T, A, C, or G. These hexagrams are decoded by using the bond-bigram decoder keys at each corner of the chart.

Consider the perfect bond-bigrams. The upper right corner holds the decoder key ⚏=^A_T for any perfect bond-bigram of evolving yin/yang. This bond-bigram's lower line codes for T; its upper line codes for A. The hexagram ䷋ just below it is the ultimate expression of its decoder key. This hexagram is 7 in binary counting but Hexagram 12 in King Wen's oracle order.

Likewise, the bottom left corner holds the decoder key ⚍=^T_A for any perfect bond-bigram of evolving yang/yin. This bond-bigram's lower line codes for A; its upper line codes for T. The hexagram ䷊ just above it is the ultimate expression of its decoder key. This hexagram symbolizes 56 in binary counting, but it is Hexagram 11 in King Wen's oracle order.

Now consider the imperfect bond-bigrams. The top left corner holds the decoder key ⚏=^G_C for any imperfect bond-bigram of static yin/yin. This bond-bigram's lower line codes for C; its upper line codes for G. Just below it sits the all-yin hexagram ䷁, the ultimate expression of its decoder key. This hexagram symbolizes 0 in binary counting, but if moved into King Wen's analog, relational order, it is called Hexagram 2.

Likewise, the bottom right corner of the chart holds the decoder key ⚌=^C_G for any imperfect bond-bigram of static yang/yang. This bond-bigram's lower line codes for G; its upper line codes for C. Just above it sits the all-yang hexagram ䷀, the ultimate expression of its decoder key. This hexagram symbolizes 63 in binary counting, but if moved into King Wen's analog, relational order, it is called Hexagram 1. In sum, all 4 decoder keys in all 4 corners decode all 64 hexagrams into all 64 DNA 6-packs.

Gazing across all 8 hexagrams on the chart's first row, abruptly we notice that *every* lower trigram in the first row is all-yin ⚏! By mirror-contrast, *every* lower trigram in the last row is all-yang ⚌! Does this mean that *every* lower yin trigram on the first row of hexagrams codes for CCC? Or that *every* lower yang trigram on the last row of hexagrams codes for GGG? Absolutely not!

Don't forget, a molecular 6-pack's identity depends upon the polarity of each bond-bigram in its hexagram. Thus in the top row of hexagrams, a lower trigram's yin ⚋ line is C *if and only if* its partner in the upper trigram is also yin ⚋. If so, then lower yin is C, and upper yin is G. They are an imperfect ^G_C pair. Likewise, in the bottom row of hexagrams, a lower trigram's yang ⚊ line is G *if and only if* its partner in the upper trigram is also yang ⚊. If so, then lower yin is G, and upper yin is C. They are an imperfect ^C_G pair.

In review, this chart correlates 64 I Ching hexagrams with 64 DNA 6-packs. By cross-coding both systems, the binary rows now throb with a vast web of analog connectivity revealed by the DNA relationships. This format exhibits the special kind of nonlinearity I call analinear. It merges binary counting, analog ratios, polarized pairs of pairs, polarized pairs of triplets, perfect and imperfect pairs.

We've now confirmed that DNA and the I Ching math are two different systems that both develop a polarized pair of pairs, which they then reshuffle into polarized 3-packs. Both systems then pair-bond those polarized 3-packs into 64 discrete 6-packs using the underlying dp-tree organization. Yes, they have variant layers of subcoding, yet they mirror a shared hierarchy. Thus both systems are two variants of the same underlying paradigm.

So now we can look at Shao Yong's binary chart from the 11th century CE, viewing its enigmatic yet harmonious order of 64 hexagrams, and since we know the hidden coding parallels, we can exclaim, "Hey, that's also the genetic code! The 64 DNA 6-packs are tucked right inside there." Later books in this series show how a universal DNA codes for pulsing beats of information in 64 co-chaos dynamics operating at the membrane interface between this universe's two huge bubbles of 3D space and 3D time. They hold the changing contents of matter and energy to manifest events in ever-emergent fractal flow.

8. Developing the decoding rationale

How did I come upon this decoding format? I followed my nose. Soon after I enrolled at the Carl Jung Institute in Zurich in 1985, I read Marie-Louise Von Franz's 1968 essay *Dialog uber den Menschen*. In it, she remarked that the genetic code and I Ching show striking parallels. After a lecture that Von Franz gave at the Jung Institute, I asked her about those parallels. She just replied that she'd not pursued the idea much past that brief mention in her essay.

So I started looking for more extended references. I found Gunther Stent's *The Coming of the Golden Age*, published in 1969. Martin Gardner explored the I Ching's binary math in *Scientific American*, January 1974. An odd, obscure pamphlet by Eleanor Morris, *Functions and Models of Modern Biochemistry in the I Ching*, was published in Taipei in 1978. I also found Martin Schonberger's *The I Ching and the Genetic Code*, published in 1980...and years later, Johnson Yan's *DNA and the I Ching*, published in 1991. Various other authors have also discussed or attempted to correlate the genetic code and I Ching.

Physician Schonberger, biologist Stent, and chemist Yan, for example, all used different cross-coding systems. Each considered various binary aspects but no analog or nonlinear aspects. None used bond-bigrams. None considered *wobble* (discussed in Chapter 7). Nor fractal aspects. Certainly not co-chaos. To me, some aspects of what they tried to do worked out, while others did not.

For instance, I think Schonberger was misled in trying to account for changing lines within the stable DNA plan, plus he used regular bigrams instead of bond-bigrams to decode DNA 6-packs into RNA codons. Those approaches did not work, and as he admitted, the result was not satisfactory or complete.

But Schonberger gave me an idea when he remarked that by arbitrarily flipping the A and G coding in his chart, parts of his result suddenly made an odd kind of sense. Schonberger wondered why that partially-flipped code had somewhat improved the result on his chart. He said although he lacked a rational justification for it, the results nevertheless "…add up to a phenomenon which simply cannot be argued away." Yet he could not explain why or fulfill it.

Schonberger couldn't explain why it half-worked, nor could he take it far enough to get both halves to work. But to me, stopping at that point was like saying if you flip-flop half the letters in the word ARTS, you get another word—RATS. Okay, ARTS and RATS are both words. But a broader view might consider other options, perhaps turning RATS into STAR. Since I agreed that Schonberger's arbitrary flip-flop gave a provocative result, I posited that half-right answers might mean something half-worked there. But what? How?

So I re-thought the parallel coding premises underlying both systems. Schonberger did not view the genetic code and I Ching as fractal variants of a larger paradigm. Eventually, I realized the most serviceable theory would use co-chaos as its basis of correlation. I employed a symbolism similar to Martin Schonberger's in *The I Ching and the Genetic Code*, but I did it in a different manner and to a different depth, allowing me to consider perfect and imperfect bond-bigrams, flip-flop code, and RNA wobble.

Gradually I developed *Decoding Steps 1* through *4* that mathematically cross-code DNA 6-packs with hexagram 6-packs. *Decoding Steps 5* and *6* are ahead in Chapter 7. They correlate amino acid actions with I Ching directives. For instance, the *wobble* that occurs in translating an RNA codon correlates with the *wobble* that exists in Line 6 of a hexagram!

By this means, I proved to myself that both systems are two fractal variants of the same underlying co-chaos paradigm. They share certain principles: (1)Both systems grow on a double p-tree. (2)Both systems develop a basic foursome as a polarized pair of pairs. (3)Both systems reshuffle their polarized pairs of pairs into polarized 3-packs. (4)Both systems pair-bond their 3-packs into 6-packs. (5)Both systems use three cross-links to bond together a 6-pack. (6)Both systems develop 64 discrete 6-packs of analinear math bonding two chaos patterns across domains in ascending orders of fractal complexity to generate its co-chaos system. To me, all of this verifies that both systems meet the mathematical requirements of classification as a dynamical co-chaos system.

But wait…there's more. Both systems correlate not only in their math but also in their meanings. RNA codons make amino acids that do various tasks; hexagrams have philosophical messages that state various directives. Chapter 7 unpacks 11 RNA hexagrams into traffic codons or amino acids whose tasks actually parallel the philosophical directives of their hexagrams. Later chapters will jump below the molecular level to show that DNA's 55 atoms even parallel the He Tu map of 55 dots passed down from most ancient China!

How old are simple yin and yang? Trigrams? Hexagrams? How old is the ancient I Ching rule for perfect and imperfect bond-bigrams? No one can say with historical accuracy exactly when or where the symbols originated.

Consider Legge's thoughts on the matter: *It is a moot question who first multiplied the figures from the trigrams universally ascribed to Fu Hsi to the 64 hexagrams of the Yi [I Ching]. The more common view is that it was King Wan; but Ku Hsi, when he was questioned on the subject, rather inclined to hold that Fu Hsi had multiplied them himself, but declined to say whether he thought their names were as old as the figures themselves, or only dated from the twelfth century B.C.*

However old the hexagrams may be, and wherever they originated, ancient China knew that two trigrams pair-bond via their three bond-bigrams inside a hexagram. In the 20th century, science showed that in the DNA double helix, codons pair-bond via their three cross-linked pairs. DNA existed long before a human hand inscribed I Ching shorthand. This TOE says both systems show parallels of co-chaos patterning that point to an underlying master code whose polarized information generated dimensionality at the mobic scale.

Most important to me is a premise left unaddressed in any previous literature I have seen: the co-chaos paradigm itself. It first appeared in a 1985 dream that I had of the universe generating itself. I *saw/was* it happening in that dream.

After years of study, I realized how to explain that I Ching math can describe a co-chaos pattern underlying the genetic code. Both systems are fractal variants templated off a far older master code of information that generated the universe itself. This TOE says that co-chaos underlies the I Ching math, the genetic code, and before that, the master code that generated our universe.

A polarized pair of pairs heralds the paradigm. Its quartet may express as 4 base molecules that bond two 3-packs into a DNA 6-pack. Or as 4 bond-bigrams that bond two trigrams into a hexagram 6-pack. Or as 4 primals (space, time, matter, energy) that bond two 3-packs (the 3D space bubble and the 3D time bubble) into this Double Bubble universe. All its fractal variants fit inside one another like Russian nesting dolls, and somewhere inside there is us.

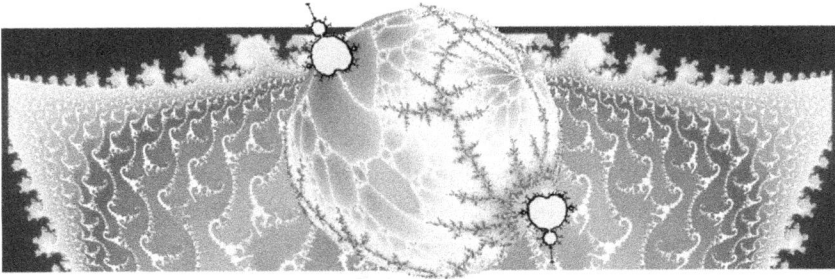

Chapter 6: Bridging Mind and Matter

1. Reconciling with fractal static

The West's linear-lauding society has sometimes not known how to absorb those who vibrate to a more analog reality. People who follow inspiration beyond logic's ability to explain may win big or lose all. They are often culture's artists, inventors, oddballs, mavericks, or misfits who march to a different drummer.

True, sometimes an eccentric talent or genius is embraced as admirable or lovable—witness the dandelion-maned, sockless, violin-playing Einstein. He developed theories on spacetime, gravitation, light, and energy that ushered in new physics emerging in the 20th century.

Changing cultural attitudes can launch new sciences able to model aspects of reality that formerly lay hidden beyond conscious attention. For instance, in the 1950s, IBM was troubled by random noise in the telephone lines carrying information to its company computers. IBM's engineers tried to drown out the phone-line static by increasing the signal strength. It seemed like a smart approach of asserting typically yang, take-charge, can-do authority. Overpower that noise. Win with power. Conquer the problem. Blot it out...

...but it did not work. IBM finally hired Benoit Mandelbrot to try to deal with the problem. He chose not to try to overpower the static. Instead, he studied the static in graphs to discover its characteristics. He found the static was consistently present and also paradoxically unusual in its finer details. Mandelbrot realized that each burst of static actually contained within it bursts of clear signal. Odder still, those periods of noise and clean transmission remained constant in ratio, no matter what scale of time he used to plot it.

Mandelbrot eventually realized that the transmitting static had a nonlinear dynamic that carried hidden relationships cycling in patterns that seemed to him much like the Cantor set continuum, which can be drawn (partially) on paper. He decided the static's dimensionality was effectively partial or fractional. By this oblique means, Mandelbrot found fractal patterns hidden within the electrical energy of so-called random static. That discovery caused Mandelbrot to invent the word *fractal* by omitting a syllable in *fractional.*

Mandelbrot also realized that the patterns of static were occurring not in space but in *time*! Or rather, in timing…in the timing of electrical bursts of patterned static running along the telephone lines. He decided to work out a new way to deal with this peculiar static by accepting it and embracing it.

How? By cycling redundancy into the information that IBM transmitted. Unlike the typical linear, binary mindset of engineers insistent on a *win/lose* fight to the death, Mandelbrot, rather than fighting the static, instead simply iterated the desired message more often to reinforce its data.

2. Reconciling with polarized life

Life itself has a shadow side that is staticky, irritating, ugly, sad…and often unspoken. Pain, hunger, illness, betrayal, loss, and death do exist. We cannot end the shadow side of things merely by denying that it exists. If dismissed, denied, devalued, those hidden truths will keep on erupting in strangely hostile ways with upped signal strength.

Shadow manifests in life's painful scenarios enacted as events in energy and matter over space and time…like that disruptive static on telephone lines carrying data to IBM's computers. Their engineers at first tried to stop it, end it, kill it, with disappointing results. Talk about a fiend that wouldn't die!

But if you can handle a disruption inclusively and well, as Mandelbrot did, what was shadowy can become creative instead of disruptive. Carl Jung said the times of heavy human shadow, if handled well, can turn its lead to gold. Heavy shadow is a potential source of creative riches if you muster enough savvy to face it, examine it, and process its unintegrated truth and potential, transforming its new vein of perception and understanding into creative gold. True alchemy. King Wen did that when he consoled his sorrow by writing down the I Ching.

A philosophical divide existed historically between the *either-or* tendency in the modern West and the *both-and* tendency in the ancient East. It's almost as if during humanity's early geographical isolation, the global culture of Earth bifurcated like a giant brain into two hemispheres. During that geographical isolation, the East went more right-brained, analog, and holistically oriented, as the West meanwhile went more left-brained, linear, and data-driven.

But things are changing now. Western culture is beginning to realize that we cannot get rid of a polarized tension merely by denying one pole exists, nor by deriding it, alienating it, fighting it, trying to push it back into the shadows. The emerging global society is shaking up old habits by stimulating cultural interaction between the globe's two hemispheres. Now, instead of acting like a split brain schizophrenically warring against itself, our global culture is starting to explore the many regional cultures as complementary aspects that bring forth new exchanges, benefits, and appreciations for humanity.

3. The I Ching can bridge mind & matter

The word *divine* can act as a noun, adjective, or verb: "The divine is divine, and it can divine." All its forms spring from a Latin root spelled *deus* or *divus*, meaning God. By learning to divine with the I Ching, you do more than wage a battle against the great unknown. You can apply its powerful imagery to your own life's issues and thus broaden your spectrum's range of personal insight.

For instance, a friend may ask, "How was your day? Was it okay?" and your mind no longer just responds with it was good or bad, light or dark, lousy or okay. Rather, your perception takes on greater tints and nuances of tone.

You may sense…oh yes, that Hexagram 62 from this morning was correct…today has not been not just good or bad, but more like this: "Wow, all day has been like hearing thunder resound and echo in the high mountains. That shocking boom of my supervisor's voice, the rocky wall of opposition that my partner threw up during our meeting…every peak and canyon in today's ups and downs kept echoing back, startling me. So Hexagram 62 alerted me by its call for vigilance, echoing in each event: 'Pay heed to the tiny details!'… So I was conscientious about the little details today, and it paid off."

What you perceived in your day's events set the lower trigram of Mountain ☶ beneath the upper trigram of Thunder ☳ . Those two trigrams form the dynamic of Hexagram 62, ䷽ *Attention to Detail.* Each trigram holds a distinct chaos pattern, and both trigrams together describe a co-chaos pattern whose dynamic is shorthanded by that hexagram.

Seeing the co-chaos dynamics in the I Ching, in the genetic code, in your own life, will weave your past and future together and give more meaning to the present. The events of your life's co-chaos patterns appear in the space-time mesh that holds emergent matter and energy on scales large and small.

Your own life is an example of this fractal process in action. Your life tends to iterate its many familiar patterns that, however, evolve their own unique details in each day's specific contents. No day is quite like another. No matter how boring a routine may seem, each emerging moment is born anew.

And a strange attractor may disrupt it. This fractal organization revealed by the I Ching fascinated its early priest-scholars precisely because it tracked life's dynamics in a pragmatic guide of conduct somehow embedded in nature itself.

What nonsense, a Westerner may scoff. How can an ancient oracle guide us through the maze of life? Impossible! Yet cultures worldwide have agreed that finally, there must be some universal root where everything in the universe—the differentiations of all things—becomes one. That oneness has taken on many names, but they all suggest innate wisdom in holistic unity.

Douglas Hofstadter asked in his preface to *The Genetic Code: Arbitrary?*,

"Can cooperation and even a seeming morality emerge purely as a consequence of the laws that govern self-replication and the universe's impersonal preference for various states?" Is there an underlying preference that is built by nature into our feelings, thoughts, attitudes, conduct, laws, ethics, morality? Is there some natural organizing factor beyond mere "survival of the fittest?"

Indeed, what does "the fittest" even actually mean in all cases? How does one live most appropriately in tune with the universe's impersonal preference for various states? What are those specific states, conditions, dynamics…and when do they change? Ancient China took such questions seriously. Its answer was to study and follow the way of the Tao. Its cosmology sets human nature and universal nature into a single system resonating with the I Ching, a text combining analinear mathematics with woolly, intuitive insight…and all of it is signaled in a spare shorthand from long ago.

4. Beyond complexity to simplicity

The best science is simple and beautiful. Newton said, "Truth is ever to be found in simplicity, and not in the multiplicity and confusion of things." The multitudinous details of our ever-emergent fractal universe are generated by a few simple rules of co-chaos that are set into motion by its dynamical system.

It is humbling to think that a tribe in ancient China somehow knew this system's mathematics so long ago. How? Intuition? Meditation? An alien visitor? I have made many speculations. Something guided them to a truth beyond the ordinary human senses. Prisoner Ji Chang first wrote down the I Ching's analinear math. It was valued so highly that eventually Ji Chang's victorious son bestowed on his father the posthumous title of Scholarly King.

The I Ching holds the essence of Taoism. It was handed down as 64 hexagrams with no technical explanations of how yin/yang poles mesh at the corners of triangles, tetrahedrons, and cubes into a lattice of spacetime in our upper bubble, pierced by what we call the arrow of time…that is only the upper half of 8-loops cycling in a tensor network across both bubbles, with its upper half reading as time in this bubble and its lower half reading as space in the conjoined bubble, a mirror-twin hidden below the mobic scale.

I have seen maps of an ancient Taoist ritual that may demonstrate a nonverbal record of this knowledge. I wonder if a ritual from 2,300 BCE called the *Yubu* (Yu's paces) is referencing the Lo Shu? I suspect it may not just be a parody of the walk of ruler Yu with his deformed leg, as some conjecture, but instead, a profound dance of the human body mimicking the procession of polarized beats that generated the universal body. It might be an attempt to dance out the progression of polarized beats that generated dimensionality at the mobic scale, far smaller than the quantum scale.

We cannot enter that level of reality with physical tools. But because of a repetitive dream, and before I'd read about the "paces of Yu," I often saw dancing in some subliminal zone of my mind's eye. It was a strange ballet of yin and yang dancing in a polarized rhythm on an ultra-tiny stage.

Current science is beginning to explore the possibilities of polarized latticing at the quantum scale. This study of lattices via "seized-up polarity" is a necessary stage for improving our knowledge and techniques. How do we study polarized latticing at the quantum scale? We perform rather elaborate, clumsy lab experiments with any polarized matter that will obligingly "seize up" enough to exhibit its frozen lattices. These lattices may appear in crystals, in metals, in the semi-metal, semi-transparency that we call graphene, or even in the impossibly slow, nearly liquid, almost transparent flow we call glass.

But even doing that much with lattices is quite difficult for us. We enter a tiny scale where all becomes relative, dependent upon what surrounds it. For instance, the study of spin glass lattices was first approached by adding a few stray atoms of foreign matter to a perfect crystal. But such fastidious work proved to be quite difficult, almost impossible. Daniel Stein described it as "attempting to study a clean mud puddle." His remark epitomizes the difficulty in staying cleanly linear during the study of something inherently nonlinear.

Is there an easier way to study lattices, perhaps even the dimensional latticing of universal space and time? I think it is possible to set up ways to study co-chaos theory in dimensionality itself, perhaps with more results than seen thus far in string theory's study of matter and energy. Applying the co-chaos paradigm to string theory might even help resolve its current impasse where the five most plausible string theory versions are now umbrellaed under M-theory's system of 8×8 factors…echoing DNA, echoing the I Ching.

M-theory was introduced in 1995 by Edward Witten. He combined five different string theories with 11D supergravity in an early attempt to unify general relativity and quantum mechanics. Witten predicted that all six theories, string or supergravity, were connected because they were all approximations of some underlying theory that was not yet known.

In later volumes, and especially in Volume 5, *Stone Soup Universe*, I explore some possibilities of how our 11D universe used co-chaos to generate dimensional latticing at the interface between the symbiotic twin bubbles of 3D space and 3D time, and how this may be shorthanded by I Ching math.

Some aspects of modern science may be trying to reinvent the wheel that the I Ching already found by tuning into a truth beyond words, but not beyond numbers. The scientific literature I've come across on matrix mechanics, on quantum lattices, on E_8 Lie Groups, on string theory, on M-theory...they often lead me to wonder if science would be better served by using the more succinct, accurate, and versatile math symbols of the I Ching in the co-chaos paradigm. Its shorthand is far simpler, yet it can describe complex, nonlinear, even analinear dynamics.

Those ancient people somehow discovered the cosmic fractal patterning of the master code hidden at a foundational level of the universal body and mind. A variant of that code generated our own bodies and minds.

Perhaps the ancients became attuned to information rising from the collective unconscious. That deeper knowing is often drowned out today by the hectic pressures and physical onslaught of modern-day life, rife with schedules so busy and full that we have no time to listen to the silence holding the Tao. Yet it is possible to vibrate with an awareness that our minds are subsets of the universal mind, just as our bodies are subsets of the universal body.

If the master code underlying nature is so simple, why haven't we noticed it before? Some people apparently did long ago, perhaps due to a dream. Which reminds me. Last night I had a dream. Most of it I cannot even remember now. I only know it was a lesson in life's elementary math, but far too colossal for me to grasp, so overwhelming that it was frightening. \

Fortunately, near the end came a quick summary of how numbers generate everything. Inside the dream, the dream-me exclaimed, *"Oh, look! Now I am not frightened. Father, you are the prime mover delineating my path. Mother, you are the double bell ringing me into life and resonating me home."*

Hey, I don't talk like that in real life. Just in dreams. Upon waking, I could remember nothing more than those odd words and something about viewing my life as wordless number play. But it was not random, accidental numbers relentlessly going nowhere. It was analinear numbers driving my life home to all meaning, all cause and effect, all resonances vibrating together in ways that I could not see or understand, but still could trust were there.

What I saw in a dream landscape beyond imagining suggests that we are something more than matter and energy constructs living in our space and time parameters. You are more than that. I am. We are pieces of divinity made small and vulnerable and trying to live together, instead of huge and eternal and lonely, trying to emerge out of an isolation of solitary invisibility.

So it seems to me that in some liminal zone where work and play, pain and pleasure, end and beginning all become the same one thing, life must be leading us all to the same conclusion and beyond.

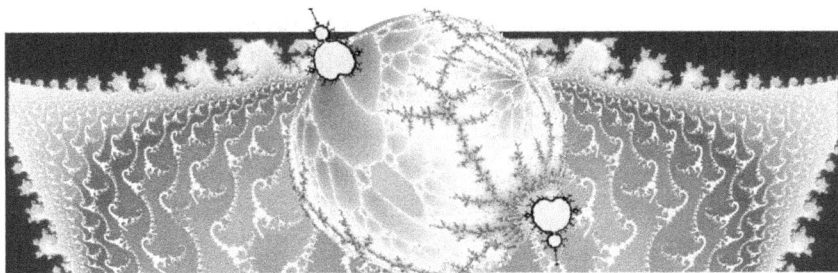

Chapter 7: RNA Builder & I Ching Correlations
1. DNA held the plan; now RNA does the building

In Chapter 1, DNA's double helix split into two strands. Each strand attracted along its length a new strand that replicated the absent DNA parent, but substituting U for T. You saw it break away from its parental template and wander off as a messenger RNA strand. Next, a small ribosome (protein-making factory) approached the mRNA strand, boarded its string of molecules and chunked them into triplets whose amino acids could produce the strands of proteins that build your body.

In Chapter 5, you saw how the I Ching's 64 hexagram 6-packs can shorthand DNA's 64 molecular 6-packs by using bond-bigrams as decoders.

Now in Chapter 7, the puzzle of genetic decoding resolves before your eyes. Here you'll find that the I Ching shorthand's decoding process automatically sorts the 64 RNA codons into 20 standard amino acid families on a chart that's ordered more sensibly than what appears in many biology texts, for it ranks all 3 *Stop* codons in a row of increasing "stop power." Then you'll see how the tasks of 11 RNA amino acids parallel their hexagrams' philosophical directives.

Is so much possible? Yes, it is. The co-chaos paradigm's consistency resides in its message, not its recording medium. The message itself— *"co-chaos patterning at work"*—can appear in any medium that lets a polarized pair of pairs organize by pair-bonded triplets into 8 × 8 = 64 polarized 6-packs.

This co-chaos message appears in the I Ching hexagram's 6-pack; it appears in DNA's molecular 6-pack; it appears in the quark's flavor 6-pack; it appears in the lepton's particle 6-pack; this series says it even appears in the information 6-packs that hold together the other bubble's 3D time and our bubble's 3D space. More broken symmetries appear up here as triplets and octaves in matter and energy as triads and octaves of elements in chemistry, as triads of particles and octaves of light in physics, as triads and octaves of sound in music.

The co-chaos paradigm begins in numbers. It uses 0 and 1 in linear, binary sequences while it also uses -1 and +1 in polarized, analog mode. The short run of numbers from 2 to 8 can merge both modes into the analinear co-chaos dynamic.

2. 128 RNA codons = 128 trigrams

The result is this organic dynamo of a chart! RNA's codon 3-packs are cross-correlated with Shao Yong's trigram 3-packs in vertical and horizontal binary order.

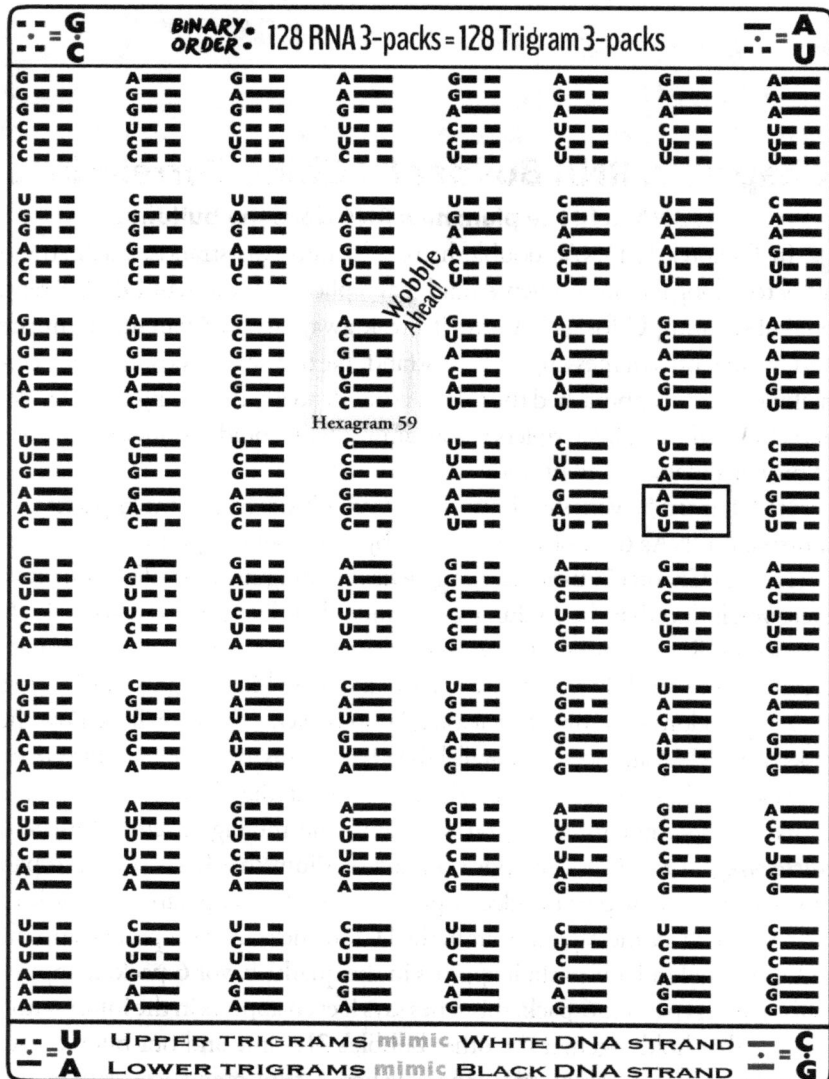

BINARY ORDER: 128 RNA 3-packs = 128 Trigram 3-packs

The 128 RNA codons correlated with the 128 trigrams

UPPER TRIGRAMS mimic WHITE DNA STRAND
LOWER TRIGRAMS mimic BLACK DNA STRAND

You may say, "Wait! I saw this same binary hexagram order back in Chapter 5, but back there, you said it correlated DNA 6-packs with hexagram 6-packs."

Yes, that's true. But now observe, there are no T's in this version, only U's. DNA uses T's, but RNA uses U's. So on this chart, view each hexagram as two separate trigrams. Each trigram stands for an RNA codon 3-pack. So this chart doesn't code for both parental strands bonded into the DNA double helix. Instead, this chart codes for two separate strings of RNA codons. Lower trigrams/codons mimic the black parental strand, but using U's, not T's. Upper trigrams/codons mimic the white parental strand, but using U's, not T's.

To maintain clarity here, we'll work only with the lower trigrams/codons that mimic the black parental DNA strand. That's why halfway down the chart, a black box surrounds only the UGA/lower trigram 3-pack. This UGA 3-pack is yoked to the lower trigram 3-pack of ☴ Wood-Wind sitting beside it. (Recall, read I Ching lines and their genetic alphabet from the bottom upward.)

Take a look on either side of that boxed UGA/☴ Wood-Wind 3-pack. Its codon neighbors are UAG and UGG. They both are also yoked to lower trigrams of ☴ Wood-Wind. In fact, that whole row of lower trigrams is 8 ☴ Wood-Winds identically populated along it! Yet their codon alphabet varies!

What! Why? It's a coding artifact left over from Chapter 5. Recall, in Chapter 5, we used bond-bigrams to shorthand DNA's 64 molecular 6-packs into all 64 hexagrams in binary order. Now in this chart, bond-bigrams still invisibly link each lower trigram to its upper trigram. So if we separated all those lower trigrams from their genetic alphabet, we'd lose part of the chart's coding data. However, we've already decoded those hexagrams, so we can safely copy all the genetic alphabet of those lower trigrams onto a new chart and not lose anything!

3. Puzzle: the recursive frame of codons

We unyoke all those lower codons that mimic the black parental strand from their trigrams and move only their alphabet code onto this new chart below. At first glance, the result doesn't look very organized…but it is. The ebook's color-coding emphasizes a pattern that isn't quite so evident in black print.

From this chart's 4 corners, C, U, A, and G distribute diagonally inward on it. On the upper left, C trickles down on a diagonal to G on the lower right. Meanwhile, on the upper right, U trickles down diagonally to A on the lower left.

CCC	CCU	CUC	CUU	UCC	UCU	UUC	UUU
CCA	CCG	CUA	CUG	UCA	UCG	UUA	UUG
CAC	CAU	CGC	CGU	UAC	UAU	UGC	UGU
CAA	CAG	CGA	CGG	UAA	UAG	UGA	UGG
ACC	ACU	AUC	AUU	GCC	GCU	GUC	GUU
ACA	ACG	AUA	AUG	GCA	GCG	GUA	GUG
AAC	AAU	AGC	AGU	GAC	GAU	GGC	GGU
AAA	AAG	AGA	AGG	GAA	GAG	GGA	GGG

The black lower strand's 64 codons

The resulting pattern is the 64 molecular triplets of RNA codons. The chart is replicated below, where you can see that along the chart's outer edge sit 28 codons. Each side counts in binary sequencing, but instead of using 0 and 1, it uses two symbols taken from the C, U, A, and G data bank.

Moreover, each side of the RNA codons chart not only counts in binary from 0 to 7, but it also is simultaneously counting backward from 7 to 0…in a coded, recursive, Escher-like run along the sides and around the whole table!

TRIGRAMS	DECIMAL	BINARY		RNA Codons							
☷	= 0 =	000	**P**	CCC	CCU	CUC	CUU	UCC	UCU	UUC	UUU
	= 1 =	001	**U**	CCA	CCG	CUA	CUG	UCA	UCG	UUA	UUG
	= 2 =	010	**Z Z**	CAC	CAU	CGC	CGU	UAC	UAU	UGC	UGU
	= 3 =	011	**L**	CAA	CAG	CGA	CGG	UAA	UAG	UGA	UGG
	= 4 =	100	**E**	ACC	ACU	AUC	AUU	GCC	GCU	GUC	GUU
	= 5 =	101	**C**	ACA	ACG	AUA	AUG	GCA	GCG	GUA	GUG
	= 6 =	110	**U**	AAC	AAU	AGC	AGU	GAC	GAU	GGC	GGU
☰	= 7 =	111	**E**	AAA	AAG	AGA	AGG	GAA	GAG	GGA	GGG

RNA Codons = Trigrams as binary & decimal numbers

PUZZLE: Can you decode the recursive border into its binary format? Recall, 0-1 can be coded by any pair of symbols. Maybe yin-yang? Or a pair of alphabet letters? The *Answer Key* to this puzzle is in Section 17 of this chapter.

4. Unpacking lower trigrams into their RNA task hexagrams

Due to the clever yet simple math, all 64 RNA codons/lower trigrams can now be unpacked to reveal their full-blown RNA task hexagrams, already automatically sorted into the 20 amino acid families that make your body. Further, their amino acid tasks reflect their RNA task hexagrams' philosophical meanings, couched as verbal directives. What! Is so much even possible?

Yes! I'll show you how. We'll do it in just two more decoding steps.

• *Decoder Step 5: Regular bigrams unpack a codon into its RNA task hexagram.*

Now we unpack each RNA 3-pack/lower trigram to reveal its full-blown RNA task hexagram with its philosophical verbal directive. To do so, we'll use 4 bigrams again, but this time, not the *bond-bigrams*. Instead, we must use the *regular bigrams* that grow on a p-tree. This code says for any RNA codon…

U unpacks a yin line into a bigram of stable yin ▬▬ ▬▬.

A unpacks a yang line into a bigram of stable yang ▬▬▬▬.

C unpacks a yin line into a bigram of yang changing to yin ▬▬▬ ▬.

G unpacks a yang line into a bigram of yin changing to yang ▬▬ ▬▬.

This unpacks all 64 RNA lower codons into their 64 RNA task hexagrams. Thus…

STABLE BIGRAMS	CHANGING BIGRAMS
U *unpacks* ▪▪ *into* ⚏	**C** *unpacks* ▪▪ *into* ⚎
A *unpacks* ▬ *into* ⚌	**G** *unpacks* ▬ *into* ⚍

This key unpacks RNA lower trigams to reveal their RNA task hexagrams

Here is the result of decoding the 64 RNA codons on the previous page…

RNA TASK ORDER: 64 RNA Task Hexagrams & Binary Equivalent

PROLINE 42 40 · LEUCINE 34 32 · SERINE 10 8 · PHENYLALANINE 2 0

43 41 · 35 33 · 11 9 · LEUCINE 3 1

HISTIDINE 46 44 · ARGININE 38 36 · TYROSINE 14 12 · CYSTEINE 6 4

GLUTAMINE 47 45 · 39 37 · 15 STOPS Ochre · 13 STOPS Amber · 7 STOPS Opal · TRYPTOPHAN 5

THREONINE 58 56 · ISOLEUCINE 50 48 · ALANINE 26 24 · VALINE 18 16

59 57 · 51 49 START · 27 25 · 19 17

ASPARAGINE 62 60 · SERINE 54 52 · ASPARTIC ACID 30 28 · GLYCINE 22 20

LYSINE 63 61 · ARGININE 55 53 · GLUTAMIC ACID 31 29 · 23 21

RNA task hexagrams automatically group into their amino acid families

Ordinary bigrams from the p-tree unpack RNA codons into their RNA task hexagrams, and the decoding process automatically sorts them into their natural amino acid families of Serine, Valine, Lysine, etc.

5. Stop codons say, "Ribosome, cease & desist!"

Each hexagram's verbal text holds a philosophical directive that aligns with its amino acid task. You may be thinking, "Wait, an automatic cross-match between amino acid tasks and their hexagram meanings cannot possibly work!"

But yes, it does. I think it's easiest for the codon novice or I Ching newbie to recognize correlations by examining first the 4 traffic codons. Traffic codons operate at a meta-level that tells a ribosome when to start or stop work, overriding its sub-routines to make proteins. This operates something like the meta-level that signals a computer when to turn on or off, overriding any individual apps with subordinated directives in their own smaller programs.

All 3 *Stop* codons sit on the chart's fourth row, sorted into a sequential trio located in the horizontal *STOPS!* box. They all tell a ribosome to stop making proteins, but they do it to different degrees and with different cessation styles. Each traffic codon's specific task echoes its hexagram's philosophical directive.

The strongest stop command is UGA. It sits at the far right with an extra black box around it. As the most powerful command, UGA is the *Full Stop* (also called *Opal Stop*). It orders a ribosome to stop work. This cease-and-desist order halts a ribosome like a period halts a sentence. In genetics, the mnemonic for UGA is "U Go Away." Below, you see UGA unpacked into Hexagram 12.

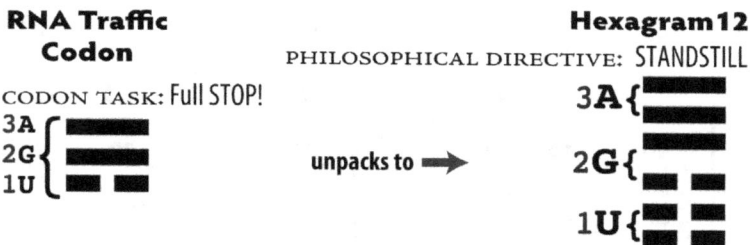

Unpacking UGA to reveal its RNA task hexagram - Hexagram 12

On the graphic, start at the bottom left to unpack UGA: **Line 1U ▬ ▬** uses the bigram of stable yin ▬▬ ▬▬ to unpack the bottom two lines of the RNA task hexagram on the right. Then **Line 2 ▬▬▬ G** on the left uses the bigram of changing yin ▬▬▬ to unpack the next two lines of the hexagram on the right. Finally, **Line 3 ▬▬▬ A** on the left uses the bigram of stable yang ▬▬▬ to unpack the hexagram's top two lines on the right. This reveals the full-blown RNA task hexagram for **UGA**. It is Hexagram 12 ▤, *Standstill*.

How well does the task of UGA, the *Full Stop* codon, fit with the philosophical directive of Hexagram 12, *Standstill?* You be the judge. Hexagram 12 ䷋ *Standstill* designates the most emphatic halt in all 64 hexagrams. It indicates that progress comes to a halt; movement does not happen. The ancient Chinese text says when the trigram of Heaven ☰ sits over the trigram of Earth ☷, they are out of relationship.

From the Wilhelm/Baynes translation: "Heaven is above, drawing farther and farther away, while the earth below sinks farther into the depths. The creative powers are not in relation. It is a time of standstill and decline.... Heaven and earth are out of communion and all things are benumbed."

Thus UGA's *Stop* task parallels Hexagram 12's directive.

Now let's consider the middle traffic codon in the chart's *STOPS* box: UAG, the *Amber Stop*. (In German, *bernstein* means *amber*. This codon's name came as a reward to co-discoverer/grad student Harris Bernstein.) UAG directs a premature termination of coding. It suppresses action, much as a yellow traffic light suppresses action, warning you to travel carefully on entering an intersection or get broadsided. In genetics, the mnemonic for UAG is "U Are Gone."

The UAG codon unpacks into Hexagram 56 ䷷ . Its ancient Chinese name translates as *Traveling Carefully, Go With Caution, Wandering, Straying from One's Element, Homeless,* or *Stranger in a Strange Land...*the title of Robert Heinlein's best-known science-fiction book. This hexagram warns you to travel carefully when heading into iffy, unknown territory, due to possible danger. Thus UAG's suppressive task parallels Hexagram 56's *caution* directive.

The remaining *Stop* codon is UAA, the *Ochre Stop*. Think of it as a sign telling the ribosome to detour away, much as police tape across a door warns people to alter their course. In genetics, the mnemonic for UAA is "U Are Away." UAA unpacks into Hexagram 33 ䷠ . It has English names such as *Retreat, Retire, Withdrawal, Step Back,* and even *Save your Bacon.* It directs you to retreat prudently to save face, honor, even skin. This UAA *Stop* redirection task parallels Hexagram 33's directive advising you to retreat.

Thus all 3 *Stop* codons have tasks that parallel their hexagram meanings.

6. Met-Start says, "Ribosome, start! But start small!"

One more traffic codon remains: AUG. It sits in the vertical box two rows below the 3 *Stop* codons. AUG is not a *Stop* command; on the contrary, it is the sole *Start* command. AUG tells the ribosome: "Start processing! Chunk this wandering mRNA strand into codons...that will become amino acids... that will synthesize into proteins." Since AUG can also make an amino acid,

Methionine, this traffic codon is often called the *Met-Start* codon.

How well does the *Met-Start* task parallel its philosophical directive? It unpacks into Hexagram 41 ☰☰, which in English translates as *Decrease, Diminution of Excesses, Start Small, Compensating Sacrifice, Dynamic Balance.* More than most hexagrams, the meaning of Hexagram 41 is rather hard for a modern, non-feudal, non-agrarian mind to grasp, as Legge pointed out: "The interpretation of this hexagram is encompassed with great difficulties."

If the I Ching's ancient cultural metaphors are hard to fathom nowadays— for example, in Hexagrams 7, 9, 41, or 47—you can read comments by various scholars through time. But the original text sometimes gets so padded over with opinions at the price of accuracy that it loses the short, sharp shock of recognition. Its succinct original text, when obscured by new verbiage, may take on different tones and directions...related, to be sure, for analogs always relate, but sometimes not quite ringing out the old truth. Confucian commentaries, in my opinion, often spin a flowery image way out past its tensile strength.

In English translations, I like Legge's terse *Judgment* for Hexagram 41. It sticks close to the original Chinese text: "In (what is denoted by) *Sun* [Chinese name for Hexagram 41], if there be sincerity (in him who employs it), there will be great good fortune:—freedom from error; firmness and correctness that can be maintained; and advantage in every movement that shall be made. In what shall this (sincerity in the exercise of Sun) be employed? (Even) in sacrifice two baskets of grain, (though there be nothing else), may be presented."

The clue to understanding this directive is the agrarian phrase "two baskets of grain." The core concept is that giving just two baskets of seed will not impoverish but instead benefit the givers by that small sacrifice. Hexagram 41 says a simple offering will bring success. By starting in a small way, the seed will produce good fortune in sustainable correctness with "advantage in every movement that shall be made."

To an ancient Chinese agrarian mind, planting seed brings a crop, and children symbolize abundance. In that rough, feudal time, conceiving children ensured the family's success. So view those two small containers of seed as parental sperm and egg. Life starts out microscopically small as male and female contribute what leads to burgeoning growth. Thus, the task of the AUG *Start* codon correlates with the Hexagram 41 ☰☰ directive. To me, it is described by names such as *Begin Small, Start Small to Succeed,* or *Small Loss for Later Gain.*

For many years I have kept careful records of oracle answers correlated with events. Experience has shown me that Hexagram 41 ☰☰ signals a time to decrease extraneous motion and redirect one's energy to start something new—a behavior, a task, an attitude, a life—but to start it small and sincerely, sacrificing a small effort for a big result. This germinating act starts something new, so this hexagram's dynamic directive aligns with the *Met-Start* codon's traffic task.

Thus all 4 traffic codons have tasks that parallel their hexagram meanings.

7. Tryptophan makes for easy going

Now let's unpack some more codons into their RNA task hexagrams to demonstrate how the math, molecules, and hexagram directives work in sync. On the chart's right border, look past the horizontal box holding all 3 *Stop* codons to find UGG. This UGG is the only codon that makes Tryptophan, and its trigram unpacks into Hexagram 35 ☷.

I call Hexagram 35 ☷ *Easy Progress*, and its dynamic directive is quite accessible to the modern mind. The Legge translation describes a feudal lord whose actions and attitude make his people feel safe, calm, and easygoing. At court, he receives rewards and is the object of favorable conversation. In other words, this hexagram describes the easy progress of a capable leader.

The two trigrams of Hexagram 35 ☷ represent the sun rising over the earth, and as the Wilhelm/Baynes translation points out: "It is therefore the symbol of rapid, easy progress…this pictures a time when a powerful feudal lord rallies the other lords around the sovereign and pledges fealty and peace. The sovereign rewards him richly and invites him to a closer intimacy." Its dynamic directive combines harmony, favor, and easy discussion.

Does this amino acid's task match its philosophical tip? Indeed it does. Tryptophan relaxes the body and calms the mind. It aids in the production of serotonin and niacin in the body, resulting in placidity. At a big holiday meal, eating at least 6 ounces of turkey plus all those carbs that help carry its bulky amino acid across the blood-brain barrier…all this tends to make a person calm and easygoing. It forestalls arguments and generates goodwill. That's why it's a good idea to serve turkey at a big holiday dinner with extended family.

I have suggested moderate doses of Tryptophan to nonpregnant clients as a natural relaxant and antidepressant. If you take it on an empty stomach near bedtime, it aids in sleeping without medication. Tryptophan is also a natural antidote for migraines. Thus the amino acid task of UGG operates much like the dynamic directive of *Easy Progress* in Hexagram 35.

8. Lysine asserts a possession in great measure

Unlike the 4 traffic codons and Tryptophan, most amino acids are made by several different codons within their family group. Since different codons unpack into different hexagrams, the math correlation remains clear, but it becomes harder to cross-match exactly an amino acid's tasks with several different philosophical directives. Yet if you bother to look, they do match up.

For instance, Lysine is made by two different codons, AAA and AAG. AAA unpacks into Hexagram 1 ☰ *Assertive Heaven*, while AAG unpacks into Hexagram 14 ☲ *Possession in Great Measure*. Can Lysine's amino acid multi-tasking truly align with *both* hexagram meanings? Let's find out.

What does Lysine do? It has powerful antiviral properties that can block the debilitating effects of cold sores, herpes, and shingles—all variants of the same virus. KAL Pharmaceuticals pointed out that taking "…extra lysine forces the virus to lie dormant in nerve cells…outbreaks are more mild when they do occur, and, if regular daily supplementation of 1 gram or so of lysine is continued, the outbreaks will occur very infrequently—perhaps never."

I have counseled some clients to take 500 milligrams to 1 gram of Lysine daily for maintenance, and up to 4 grams spread throughout a day during an outbreak. I emphasize, however, that Lysine does not kill a virus. It merely keeps the virus inactive and suppresses its symptoms. Such assertive, yang action on a virus is consistent with Hexagram 1 ☰ *Assertive Heaven*. Especially for shingles, its relief can be a possession in great measure, figuratively speaking.

Lysine also stimulates the hormone production that helps a child's arm and leg bones lengthen. Special structures near the bone ends called epiphyseal growth plates are stimulated by lysine to lengthen those bones properly so a maturing child develops an adult skeletal frame, thus avoiding dwarfism. The result is quite literally a possession in great measure. Thus this multi-tasking amino acid reflects the message of Hexagram 14 ☲ *Possession in Great Measure*.

9. Aspartic acid relieves toxic imbalance & aids stamina

Aspartic acid is another amino acid made by two codons—GAC and GAU. It is taken as a supplement mainly for two tasks: (**1**) increasing metabolic performance and endurance and (**2**) eliminating surplus ammonia after great physical exertion, thus avoiding toxicity that slows the body down.

How well does Aspartic acid's multi-tasking correlate with its two hexagram meanings? The GAU codon unpacks into Hexagram 32 ☳ . Its ancient Chinese name is translated variously as *Duration, Constancy, Long-Lasting,* or *Endurance,* so this hexagram's directive accords well with Aspartic acid's documented ability to improve athletic stamina and endurance.

Aspartic acid's other codon, GAC, unpacks into Hexagram 28 ☱ . Its name is translated variously as *Great Excess, Stress, Strain, Overload, Crisis,* or *Critical Imbalance.* In this hexagram's dynamic, a heavy burden causes an imbalance that needs remedying. Wilhelm/Baynes says, "The ridgepole sags to the breaking point. It furthers one to have somewhere to go."

The heavy burden can be likened to a surplus of ammonia after great physical exertion, bringing toxic imbalance. I myself felt this GAC effect two years before I began correlating amino acids with hexagrams. Once I hiked all day in the Swiss mountains and got under-hydrated. That night I went to bed so tired that I grumbled to myself, "Oh, I need some help!"

During the night, a short dream showed me a vague, hazy picture of stuff that I knew was toxic, and a dream voice directed, *"Get rid of the ammonia."* I woke up still tired, thinking, "What a stupid dream! Makes no sense. I don't have a bottle of ammonia!" Yet I knew the weird dream must mean something, so I did a computer search. Wow! Who knew? There's a link between exercise, hydration, toxic ammonia, and Aspartic acid? I found a pharmacy, bought Aspartic acid, took 4 tablets with water, and 45 minutes later, I felt fine.

10. Histidine in right balance turns anger to abundance

Histidine is another amino acid made by just two codons—CAC and CAU. Do its amino acid tasks parallel their hexagram meanings? Consider this:

Histidine's CAC codon unpacks into Hexagram 49 ䷰ *Revolution*. Its two trigrams of Fire ☲ under Lake ☱ burn and boil, seething with revolt. Dr. Donald Gerber's research at Downstate Medical Center-New York found that arthritics' blood contains about ¼ as much Histidine as healthy people's blood, so he tried giving them large doses of Histidine. Gerber found the results "… benefit these folks greatly. Some of them showed improvement with only one gram of Histidine daily; others needed as much as six."

How does Histidine work? It helps form myelin sheathing around nerve cells to protect them from damage, thus keeping the nerves steady. Other studies suggest that arthritic people often show a pattern of inner rigidity and repressed rage that "makes the blood boil," to quote an idiom. The correct amount of Histidine can help free up one's mind enough to discharge anger in a safe, healthy way that avoids rigid mental stiffness, instead of keeping that anger locked within the body, where it seethes silently into arthritis.

Histidine balance is important. Too little of it lessens nerve protection, depleting the brain's natural alpha rhythms so much that excitatory beta waves dominate, resulting in anger and tension…yet too much Histidine is found in the brains of schizophrenics! A Histidine-related adrenal imbalance can also affect the brain's inhibitory transmitter, causing allergies, nervous problems of deep anxiety, inflammation in the muscles, and outbursts of rage.

Histidine's other amino acid, CAU, unpacks into Hexagram 55 ䷶ *Full Abundance*. Its trigrams are Fire (here lightning) ☲ under Thunder ☳, as an abundance of beneficent natural power. Histidine in the right balance turns the angry power of Hexagram 49 ䷰ *Revolution* to a more beneficent use.

Since the human body cannot manufacture nine amino acids, including Histidine, it may be supplied by foods such as chicken, turkey, fish, soybeans, milk, cheese, nuts, seeds, whole grains, and eggs. Eating these can stop the body's turmoil and turn its energy to a more happy expression as it soothes the stomach, relieves heartburn, nausea, and heals stomach ulcers. This full and happy state parallels Hexagram 55 ䷶ *Full Abundance.* In my 70's now, I take 300 mg. of Histidine 3 times weekly to avoid stiffening up mentally and physically.

The correct balance of Histidine can even bring sexual abundance. Women who cannot achieve orgasm have taken more Histidine to good effect. Paradoxically, men who often experience premature orgasm may have too much Histidine, which can be moderated by taking Methionine and Calcium.

11. Amino acids with many codons are hard to typecast

At the other extreme of coding, 3 big families—Arginine, Leucine, and Serine—use any one of 6 different codons to make their amino acid! Arginine, for instance, uses the 6 codons of CGC, CGU, CGA, CGG, AGA, and AGG. All 6 codons make Arginine, but each codon unpacks into a different RNA hexagram with its own philosophical directive.

I won't try to convince you that all 6 hexagram meanings apply equally well to the multi-tasking Arginine family of codons…even though, truth be told, I can think of circumstances in which each might apply, but it is admittedly a subjective judgment call. At this point, since the RNA tasks have become multi-pronged and tenuously overlapping, I won't correlate all 6 different codons of the Arginine family with their 6 hexagram meanings.

Nevertheless, perhaps enough correlation has been shown to suggest that amino acid tasks and their correlated hexagram meanings appear to dovetail to a degree well beyond chance, at least for one familiar with amino acids and the I Ching. As mentioned, it is easiest for a newcomer to see a clear correlation between the 4 traffic codons and their hexagram directives…and also for any amino acids that are coded by only one, two, or three codons.

Of course, since all 64 codons group into only 4 traffic codons and 20 standard amino acid families, how can a full correlation between amino acids and hexagram meanings even be proved scientifically? I cannot give you a binary right or wrong on this, but in the webby resonance of analogs where no one can prove me wrong—or right—I do intuit a strong correlation.

In a strange way, analog associations are not right or wrong, anyway; they are just nearer or further from a central truth, as though accuracy moves in resonant rings of connectivity that become distorted and dispersive when too far from the center. To notice analogs and heed them, you need to true up your own inner gyroscope as you walk through the echoing halls of intuition toward a central truth located somewhere in that labyrinth of right-brain knowing.

I also admit that seeking truth in that echoing labyrinth of possibility is scary for those who shun dreams, abhor myths, dislike symbolism. But those too hidebound by logic may ignore or neglect their own feelings or those of others around them. They may project their denied emotional issues onto others, trumpeting accusations, unaware they are just externalizing their own shadow rather than confronting it internally and dealing with it productively.

Learning to walk through the right brain's analog labyrinth toward a central truth can open up rich and beautiful panoramas of inspiration and creativity. Nevertheless, its mysterious richness needs to be paralleled by left-brain logic, allying logic's *either/or* summary judgment with holism's *both-and* perspective.

In decision-making, to find balance on a specific issue, learn to recognize if your left brain and right brain are working together in agreement…or do their conflicting answers set your psyche at war against itself, at odds with others?

Do you keep changing your mind about something, staying stuck on the indecisive horns of a dilemma? If so, the I Ching can coax forth new agreement between your left- and right-brain's disparate ways of knowing to open a larger reality—one that is more consistently illuminating, since better-informed expectations become more congruent with what actually happens.

Here is the nebulous place where Chinese medicine operates. From personal experience, I know it can work quite well. Once in China, I sat weak, coughing, feverish, among 30 poor Chinese peasants, all of us lined up on hard, wooden benches around the inner wall of a small room.

The doctor called me over to her desk in the center. I sat on a wooden chair. She looked long into my eyes, then at my tongue. She made 6 different pulse readings on my right wrist and 3 on my left. She diagnosed pneumonia. No lab tests, no shots, no hospital. She wrote out a prescription for me and told me how to cook it.

I walked across the hall to a window above a half-door, where I turned in my scrip to a man and watched him walk around the big herbal storeroom lined with shelves, gathering from various bins the scoops of dried blossoms, roots, leaves…dead insects! He portioned them into three brown paper bags. Three doses. I paid about $3.00 and went home.

At home, I emptied one bag onto a tray to check what was inside. The tray was full of what looked like 13 different kinds of dried twigs, blossoms, bark, shavings, leaves…and what appeared to be a huge, brown, dried cockroach!

The doctor's instructions: each night, brew up one bag of ingredients in an earthenware pot and drink its tea, then go to sleep. I did so…whereupon each night I had the most lovely, soothing, technicolor dreams of peace and joy. My pain, weakness, and mucus diminished. On the fourth day, I was well. I even continued to have those harmonious dreams every night for another week…until, oops, away from clean water, one day I drank a bottle of Coke.

Whoa! I went right back to feeling normal. And that normal was less good.

12. Codons wobble into their amino acid families

A final coding question remains. Each of the 64 RNA codons is unique, so how do they manage to tidy up into just 20 amino acid families and 4 traffic codons grouped on a chart?

Okay, yes, the *Met-Start* traffic codon does double-duty; along with its GO task, it also makes an amino acid. But normally, the other 3 *Stop* traffic codons do not. (In Stop codons, a unique synthesizing mechanism may occasionally allow rare amino acids to get into a protein.)

The question persists: what arcane twist in the process unpacks all 64 RNA codons into their 64 RNA task hexagrams, correlates their amino acid tasks with appropriate philosophical directives…and then even automatically sorts them neatly into just 20 amino acid families, plus the 4 traffic codons?

The 20 families automatically group on a chart, but they don't necessarily share a first, middle, or last name. If you compare a human name like Agatha Clarissa Christie to a codon name like ACC, you'll find that amino acid families don't all sort by the same first, middle, or last name. And any amino acid family made by 6 different codons will even subdivide to sit in 2 two different locales on a chart. Three do so—Arginine, Leucine, and Serine.

So what is going on that sorts the amino acids into just 20 families? This truly esoteric kink shared by both the genetic code and the I Ching: *wobble!*

• *Decoder Step 6: Wobble occurs in the genetic code and in the I Ching.*

What is wobble? It is a sublime flaw built into co-chaos itself. First, consider how the trait of wobble operates in the genetic code. You already saw a ribosome "factory" chunk the RNA string of molecules into codons. Now picture this: immediately, a tRNA anticodon "worker" shows up. Its job is to verify the accuracy of each RNA codon "chunk" via its own mirror-codon.

Below left is the mRNA codon of 1C-2G-3U now being verified by the tRNA worker anticodon. Its response should be a perfectly polarized answer of 4G-5C-6A. But oops! Instead, the tRNA worker's answer is sloppy, even downright *degenerate!* Yes, that's the actual term used in genetic lingo!

Codon-
mRNA

**Anticodon-
tRNA worker**
(job-verify mRNA codons)

3U • 6 *Wobble!* A, G, & I *!!!*

2G • 5C

1C • 4G

Wobble relates to the 6th position in both genetic code & I Ching

Oh, sure, things start out fine. Position 1C meets Position 4G. They fit. Then Position 2G meets Position 5C. They also fit. So far, so good. But that third pairing—oops!—upon finding U in Position 3, the anticodon goes all wobbly in Position 6. (A telltale ring around Position 6 denotes its degenerate wobble.)

Why does the wobble happen in Position 6? Because something is amiss in Position 3. Recall, every strong T molecule in the parental DNA double helix lapsed into a weaker U molecule in its mRNA offspring's single strand.

That weaker U in Position 3 now befuddles the worker into making a bungled, *degenerate* response in Position 6. It may offer a response of A or G or even throw in a wild card of I (Inosine). By offering so many bonding options, it acts "degenerate." Moreover, this is just one of several different wobble scenarios. If that worker had answered CGU with a stalwart reply of GCA, all 6 molecules of the codon and anticodon would verify a perfect cross-match. You can spot the perfect version in *Section 2* of this chapter in the chart of *128 RNA Codons = 128 Trigrams*. A bottom codon of CGU (gold) sits under a top codon of GCA (blue)...but recall, they exist in two separate RNA strands.

13. Why does wobble work out well?

Wobble's fortunate laxity is what lets the worker anticodons sort all 61 codons that make amino acids into just 20 families. Moreover, workers will decode 6 different codons into the Serine family, yet on a chart, that Serine family is subdivided into two separate locations—as UCC, UCU, UCA, and UCG in one area—and AGC and AGU in another area. You can see this occur in Section 4 of this chapter on the chart of *RNA Task Hexagram-Family Order.*

It was Frances Crick who first identified *wobble.* He realized the verification rules become relaxed at Positions 3 and 6, allowing Position 3's molecule to pair with any of several possible molecules at Position 6. This turns wobbly Position 6 into the most generalized translation spot in the dialogue between the mRNA codon and tRNA worker, marked by that ring around Position 6.

But wobble turns out to be a happy accident. In fact, the genetic code has many happy accidents, suggesting that sometimes they are not errors at all, but rather, a careful winnowing by evolution. Recall another, earlier "mistake" where

RNA mis-mimicked T as U? That laxity is what freed the RNA strand to roam and deliver DNA's building plan to the ribosome "factories" that make us all.

Likewise, the anticodon's degeneracy in Position 6 delivers a remarkable gift. That laxity, seemingly slipshod, is what consolidates all 64 codons into just 20 amino acids and 4 traffic codons. Due to wobble's degeneracy, several different codons can make the same amino acid…and thus, thankfully, help minimize the effect of an occasional mutation mistake in a gene.

14. Hexagrams have wobble, too

Does a hexagram have anything like wobble? Yes, it does, and for it also, the wobble occurs in Position 6! Below, the wobbly Position 6 anticodon offered by the tRNA worker is paralleled by Hexagram 59's wobbly 6th line.

Codon-mRNA	Anticodon-tRNA worker (job-verify mRNA codons)	Hexagram 59 Lower Trigram	Upper Trigram
3U · 6 *wobble!* A,G,&I !!!		▬ ▬3	▬▬▬ 6 *wobble!*
2G · 5C		▬▬▬2	▬▬▬ 5
1C · 4G		▬ ▬1	▬ ▬4

Wobble relates to the 6th position in both genetic code & I Ching

An old I Ching rule for the sixth line of a hexagram mimics an anticodon's genetic wobble at Position 6. Ancient texts say Line 6 comes too late in the line sequence to maintain the hexagram's strong message. It is "furthest removed from worldly influence" and is "closest to heaven." Thus the purpose of Line 6 is not readily visible at the level of normal consciousness.

Line 6 is distanced from the other hexagram lines in its position and often even from our human ability to understand its intention very well. Line 6 has a different aim. It reaches beyond an individual hexagram's dynamic to serve a higher goal that points toward the I Ching's larger perspective on reality.

As a result, according to ancient texts, Line 6 usually exhibits a meaning that seems unrelated, remote, or sometimes even deviant. Why? Because it looks beyond the scope of that single hexagram to the larger view. It reaches for a higher order of purpose that rests within the whole I Ching system.

To quote a modern source on the I Ching's hexagram meanings and philosophy, LiSe Heyboer says of Line 6…"Unworldly things, universal values and laws, religion, mysticism, revelation, everything that surpasses tangible life…Humble obedience to universal values and laws." That quotation comes from an interesting Dutch website, and I provide a link for it in the ebook version. Given the ephemerality of personal websites, I do not know how long it will work, but this beautiful site reveals the I Ching's impact in one person's life.

In sum, the "degenerate" wobble of an anticodon at Position 6 is a seeming laxity that lets several different codons become transcribed into the same amino acid, thus minimizing mutations. That genetic wobble at Position 6 reaches for a higher purpose of promoting viability in biological life itself.

Likewise, the wobble of Line 6 in a hexagram seeks higher purpose in what the ancient Chinese called the Tao, the flow of universal mind. It allows a hexagram to look beyond itself to further the big picture. This trait of wobble is yet another parallel that exists between both genetic code and I Ching.

15. The universal scope of yin & yang

You've seen how DNA's molecular 6-packs correlate with the I Ching's hexagram 6-packs. You've seen how RNA codons unpack into RNA task hexagrams that are already automatically sorted into 20 amino acid families. You've seen how amino acid tasks are reflected in their hexagram directives.

Assessing all of this requires logic, of course. Still, ordinary logic would not expect such congruence on so many levels. Why, even the hexagram directives parallel their amino acid tasks in a congruence of mind and matter…it's nearly impossible unless both systems are based on the same underlying paradigm.

The co-chaos paradigm underlies the universe itself. Its special kind of math is *not* just binary units, nor just analog resonances. It employs both, organizing numbers into a dynamical system that is nonlinear, even analinear. The linear numbers in sequence drive to a result, yes, but the analog numbers in relationship explore networks of ongoing process…for mind and matter.

The ancient Chinese phrased it this way: "What likes to go together?" The answer is not just sweet and spicy. Nor just peanut butter and jelly. Nor just caviar and champagne. It is also mind and matter. This way of conceptualizing reality is very different from a typical linear, *either-or* view.

The I Ching can help a modern person expand beyond the limits of a strictly linear, logical mindset…mind you, not to drop the elegant binary code, logical syllogisms, and catalogs of quantities…but instead, to broaden that perspective enough to include also the fluxing networks of analog qualities in ceaseless, shifting flow.

Learning to understand and use the I Ching definitely stretches the mind's logical boundaries and opens up new vibratory fields of mental acuity. As with traditional Chinese medicine, it regards nature holistically. By studying the I Ching, one can even begin to fathom how acupuncture and Chinese medicine first set their roots in a holism that is not very accessible to our current culture, trained as it is to favor linear logic, and especially in science.

Yet a holistic, right-brained approach can be rational, and indeed, quite literally so, for it honors life's ratios of relationship continually at play, rebalancing in a shifting network that is symbolized by the tai chi ball ☯. To the ancient Chinese mind, the universal way of the Tao provided the ever-emergent flow of reality's continually evolving relationships.

16. The importance of the co-chaos paradigm

Perhaps you suppose that several other kinds of mathematical cross-coding would serve equally well to make the hexagram 6-packs automatically code for the DNA 6-packs? Would automatically sort the RNA codons into 20 family groupings? Would correlate their amino acid tasks and hexagram directives?

I doubt it. Not so far as I can tell. I've spent over 35 years now studying it. The method I've shown you—correlating DNA 6-packs and hexagram 6-packs, then cross-coding RNA codons with trigrams, then unpacking their RNA hexagram messages—all this dovetails the genetic code and I Ching to reveal the co-chaos paradigm that underlies both systems.

Along with the genetic code's rules for the transcription and translation of amino acids, it also considers the development of yin and yang on the double p-tree, the obscure bond-bigrams, wobble, and so on. What's more, it matches molecules with messages and matter with mind. The only spot that allowed me a degree of coding alternatives was a subroutine of assigning which bond-bigram to which molecular pair. There, I strove for impartiality (and humor).

Why is this co-chaos paradigm so important? Why does it matter so much to realize that both systems share the polarized pair of pairs, the pair-bonded triplets, the 64 polarized 6-packs, the polarized bifurcation tree, and even that zany wobble? By seeing how the I Ching shorthands the genetic code, both variants can offer us a Rosetta Stone with two known and one unknown code… the master code of polarized information that generated the universe itself.

Suppose you had to inscribe the co-chaos paradigm for another culture or another world. How could you possibly code anything more universally, simply, and economically than by using yin and yang? It writes out organic life's code using just two kinds of lines: ▬ ▬ and ▬▬▬. This symbolism is far more potent than the alphabet of T, C, A, and G. Its analinear math shorthand is universal.

17: Decoder Key to the puzzle

Recall that puzzle in Section 3? Here's its answer. Each alphabet letter in the chart stands for a molecule. They organize into molecular 3-packs called codons. The blue band running around the alphabet's outer edge holds 28 codons. Each gray edge counts in a hidden run of binary numbers from 0 to 7. The *Answer Key* decodes that first column of binary numbers in darker gray.

A=0 C=1	C=0 A=1	0 7	1 6	2 5	3 4	4 3	5 2	6 1	7 0		
111=7	0=000	CCC	CCU	CUC	CUU	UCC	UCU	UUC	UUU	7	0
110=6	1=001	CCA	CCG	CUA	CUG	UCA	UCG	UUA	UUG	6	1
101=5	2=010	CAC	CAU	CGC	CGU	UAC	UAU	UGC	UGU	5	2
100=4	3=011	CAA	CAG	CGA	CGG	UAA	UAG	UGA	UGG	4	3
011=3	4=100	ACC	ACU	AUC	AUU	GCC	GCU	GUC	GUU	3	4
010=2	5=101	ACA	ACG	AUA	AUG	GCA	GCG	GUA	GUG	2	5
001=1	6=110	AAC	AAU	AGC	AGU	GAC	GAU	GGC	GGU	1	6
000=0	7=111	AAA	AAG	AGA	AGG	GAA	GAG	GGA	GGG	0	7

BINARY DIGITS BINARY
Answer Key for first column

0 7	1 6	2 5	3 4	4 3	5 2	6 1	7 0

The 64 codons are framed in a recursive run of binary code

Along each outer edge, no codon holds more than two *different* letters. For instance, the left column uses only C and A. Alphabet can be used to count in binary. Say that C = 0 and A = 1. In this case, the left column of codons counts in binary from 0 to 7 by moving from *top to bottom.*

But on the other hand, if you say that A = 0 and C = 1, then instead, the column of codons counts from 0 to 7 in binary by moving from *bottom to top.* In other words, each end of the column has 3 symbols that together stand for 7, yet they also simultaneously stand for 0!

This same thing happens on every other edge, too, but just using different alphabet symbols. For instance, the top edge uses only C and U. They can symbolize either 0 and 1, or 1 and 0. Likewise, the right column uses only U and G, which can symbolize either 0 and 1, or 1 and 0. Finally, the bottom row uses only A and G, which can symbolize either 0 and 1, or 1 and 0.

The result is binary counting that runs both backward and forward along each edge of the border. The chart is polarized and mirror-symmetric in several ways, as signified by the vertical and horizontal lines that divide it into 4 quadrants. The ebook uses color-coding to make the cross-coding process clearer.

Since all 4 quadrants flip-flop their alphabet and thus their number sequencing, the resulting run of binary numbers creates a recursive boundary running around the chart's border. This recursive property on all 4 sides is reminiscent of motifs in an Escher drawing, where fishes reverse into birds, or black and white birds fly in two directions simultaneously.

I say it's time now to relax into the natural magic of the co-chaos paradigm. Instead of probing for more information, why not simply view the binary order of hexagrams as an intriguing design…and even enhance that approach by flipping those 64 hexagrams sideways to shut off the left brain's tallying habit:

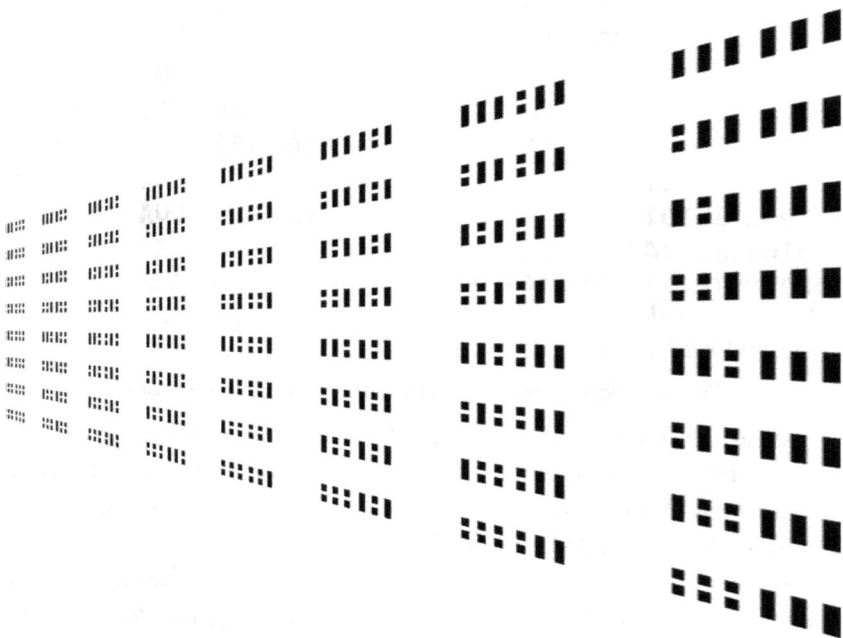

Wall of binary hexagrams

Now yin and yang turn into mere blips on a parking garage screen as we walk by. Or maybe they're dark openings in a garden wall divider. Passing by the crenellations, I marvel at the enigmatic impact of a strangely evocative design. I am finally looking at the genetic code simply as an artistic impression, so now it seems appropriate to lighten up and have some playful fun with the paradigm.

Consequently, I'm deviating from my usual plan that alternates left-brain and right-brain chapters. Next up are three right-brained chapters that play around with genetic code and I Ching math as artful designs. After that are three left-brain chapters discussing and showing how the atoms in DNA molecules parallel an ancient I Ching map in China's remote past. Then the final chapter, as usual, explores a hexagram. Since this is Volume 3, it will be Hexagram 3 ☲☷ *Laboring Birth.*

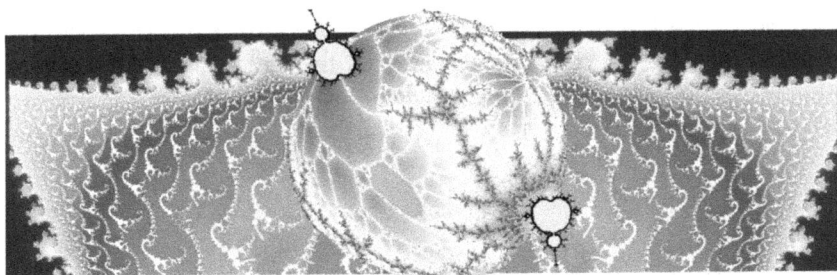

Chapter 8. DNA as Design

The next three chapters explore artful designs that nature has hidden in the math of the genetic code. I include these designs because the genetic code invites not only biological and mathematical examination but also artistic recognition. Artful designs sit everywhere in the subtext of organic life, but we usually don't consider how nature embeds designs in every living cell. In this chapter, let's noodle around in DNA's spiraling, artful, analinear ease for a bit. Relax on the parallel banks of the double helix to pick out small designs in its genetic flow. We'll watch number castles hover in the air. We'll discover the genetic code in a new way, not as a code to be broken, but as designs to be enjoyed.

1. The genetic code uses the golden ratio

The golden ratio is evident in the genetic code. This ratio is the most famous aesthetic rule in history, yet it is quite simple and perhaps so pleasing because of that simplicity. The harmonious proportions of the golden ratio let line **A** divide into the longer piece **B** and the shorter piece **C** in such a way that **A** is proportional to **B** as **B** is proportional to **C**. We can show it below as...

B | **C**
_____

A
A *is to* B *as* B *is to* C
THE WHOLE LENGTH **A** DIVIDED BY THE LONG PART **B**
EQUALS THE LONG PART **B** DIVIDED BY THE SHORT PART **C**.
A IS ≈ (ABOUT) 1.618 TIMES **B**, AND **B** IS ≈ 1.618 TIMES **C**.
THIS GOLDEN NUMBER OF ≈ 1.1618 IS CALLED PHI (φ).
The "divine ratio" of the Golden Mean

The golden ratio's beauty hinges upon using **B** as a shared reference point. This shared term **B** supplies the common ground, the middle way, the golden mean. It is what makes the proportions of the whole become so harmoniously pleasing. In nature, the golden ratio is found in our own DNA structure, a spider web, an apple blossom, a sunflower, a pine cone, a curving fern, a pineapple, an angelfish, an ant, a lily, a nautilus shell....

Below are a few black-and-white examples of how life itself carries the golden ratio. The genetic code generates all these mandalas of the golden ratio by promoting functionality along with pleasing aesthetic proportions.

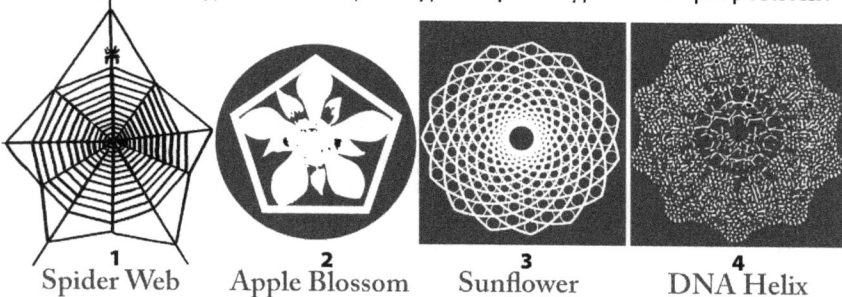

| **1**
Spider Web | **2**
Apple Blossom | **3**
Sunflower | **4**
DNA Helix |

Mandalas of the golden ratio

In Image 1, a spider constructs a soft star by spinning a segmented plane of webbing that spirals outward. Its proportions are based on the golden ratio. Image 2 cuts across an apple to reveal the mandala at its core that looks like an artsy logo mimicking the apple blossom that grew it. In Image 3, the seed pad of a sunflower reveals an intricate rendition of the golden ratio. In Image 4, you see the double helix of DNA viewed from above; you are gazing down at its DNA molecules along the spiraling ladder. Its phi proportions create the effect of a molecular rose window that is constructed on the golden ratio.

Since the golden ratio is built into nature itself, the early Greeks considered it to be what puts beauty into nature for us. They said nature is beautiful because it reveals an essential truth to us—physically, aesthetically, even morally. What truth? The Greeks said that nature shows us the middle way is desirable, beautiful, necessary, and true because it mediates a balance between two extremes...excess and deficiency.

Beauty and truth were inextricably meshed for the ancient Greeks, so they copied the golden ratio from nature and put it into their art, math, and philosophy. It appeared in their sculptures, paintings, music, and architecture. It gave their buildings the harmonizing proportions of columns and entablatures that we now term "classical."

Since the days of ancient Greece, many others have also loved the golden ratio. Fibonacci in 1202 CE extolled its basis in the golden number of 1.618 or *Phi*. Kepler praised it as the "divine proportion." Descartes charted it in the spiral of a chambered nautilus. Bernoulli investigated its logarithmic spiral extensively, finding it so heavenly that he wanted it engraved on his tombstone, along with the Latin words *Eadem mutata resurgo*—"Although changed, I shall arise the same." The golden mean is perhaps the simplest artful design in nature.

2. Bringing in the DNA Sheaves

Now we'll look for simple number rhythms in the needlepoint-like *Bringing in the DNA Sheaves* below. Each sheaf holds 64 DNA/hexagram 6-packs:

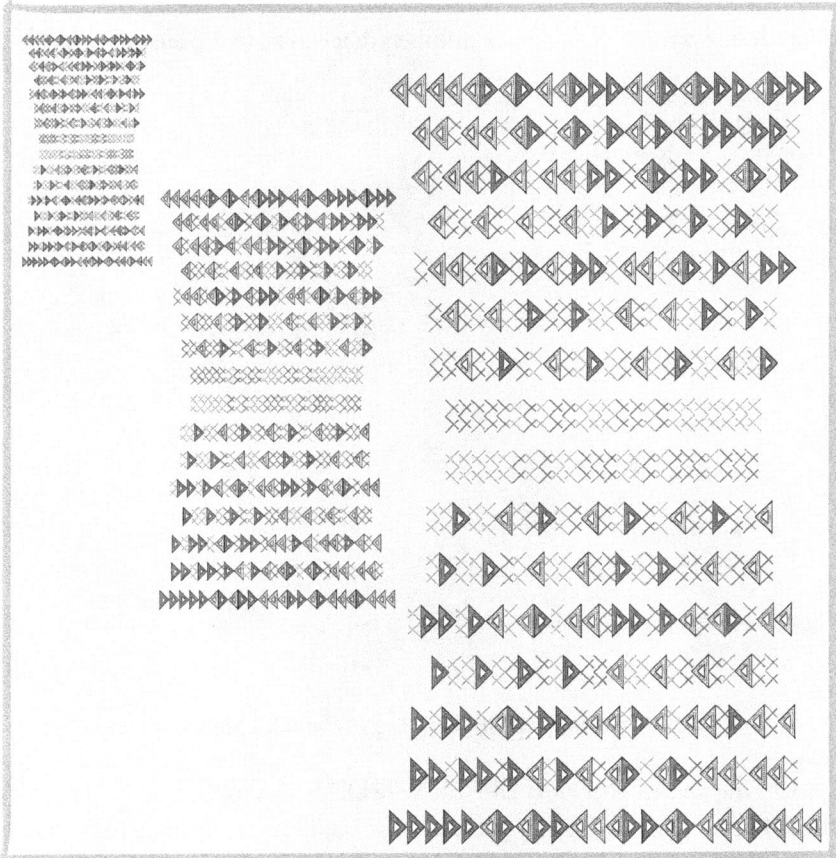

Bringing in the DNA Sheaves and its code key

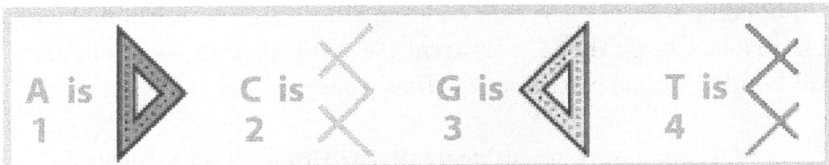

This image touches my heart for reasons that I cannot exactly describe. It reminds me of my mother, who did a lot of needlepoint in her day. It also recalls something of the distant past when sheaves of grain stood in the fields of landowners, and after the harvest, the poor were allowed to glean any stray grain stalks to stay alive. Staying alive is what DNA is all about.

3. A single sheaf in image & number

Below on the left is a sheaf of DNA. Its designs code for molecules. A decoder key sits in the center: 1-Adenine, 2-Cytosine, 3-Guanine, and 4-Thymine. The number chart on the right decodes the sheaf's designs into digits. It shows them as 3-packs of numbers that equate to 3-packs of molecules.

Top Half UPPER TRIPLETS

333	331	313	311	133	131	113	111
334	332	314	312	134	132	114	112
343	341	323	321	143	141	123	121
344	342	324	322	144	142	124	122
433	431	413	411	233	231	213	211
434	432	414	412	234	232	214	212
443	441	423	421	243	241	223	221
444	442	424	422	244	242	224	222
222	224	242	244	422	424	442	444
221	223	241	243	421	423	441	443
212	214	232	234	412	414	432	434
211	213	231	233	411	413	431	433
122	124	142	144	322	324	342	344
121	123	141	143	321	323	341	343
112	114	132	134	312	314	332	334
111	113	131	133	311	313	331	333

CODE KEY
1 is A is
2 is C is
3 is G is
4 is T is

Bottom Half LOWER TRIPLETS

A Single Sheaf

A single sheaf from Bringing in the DNA Sheaves

On the chart's **Top Half UPPER TRIPLETS**, the top left 333 equates to the GGG codon. To find its molecular bond, look on the chart's **Bottom Half LOWER TRIPLETS** for 222 that equates to the CCC codon.

The upper and lower halves of the chart sit in balance. They obey the pair-bonding rules of the genetic code, and they also obey the bond-bigram rules of the I Ching. They also represent 128 vertical period 3 windows (vp3s) pair-bonded into 64 patterns of co-chaos dynamics.

4. The sheaf reassembled into DNA/hexagram 6-packs

In the next chart, the sheaf's upper and lower halves are reassembled into 64 DNA/hexagram 6-packs sitting in binary, ancient *xiantian* order. Bindu points now connect those triplets from the sheaf's *Upper Half* with their companion triplets in the sheaf's *Lower Half* through all rows. This turns the sheaf array into 64 DNA/hexagram 6-packs, yet now they are coded in digital numbers.

DNA molecules	333	331	313	311	133	131	113	111
1 is A	222	224	242	244	422	424	442	444
	334	332	314	312	134	132	114	112
2 is C	221	223	241	243	421	423	441	443
3 is G	343	341	323	321	143	141	123	121
	212	214	232	234	412	414	432	434
4 is T	344	342	324	322	144	142	124	122
	211	213	231	233	411	413	431	433
	433	431	413	411	233	231	213	211
	122	124	142	144	322	324	342	344
	434	432	414	412	234	232	214	212
	121	123	141	143	321	323	341	343
	443	441	423	421	243	241	223	221
	112	114	132	134	312	314	332	334
Upper 3-pack ●= *bindu point* Lower 3-pack	444	442	424	422	244	242	224	222
	111	113	131	133	311	313	331	333

Viewing the binary order of 64 DNA/hexagram 6-packs as digits

The result puts all 64 DNA/hexagram 6-packs into binary order, yet they're now symbolized in a digital number format that no longer looks like binary counting. This digital setup makes it easier to examine some of the system's relational bonds.

5. The ubiquitous 15

The distribution of numbers in this chart reveals an awesome symmetry of number relationships. You may be as startled as I was upon stumbling into some of its finer details. For example, in each DNA/hexagram 6-pack, the bindu point (•) between its two triplets bonds an interesting internal relationship. Spot the chart's lowest 6-pack on the left. Consider its three bond-pairs of 4 • 1, 4 •1, 4 • 1.

```
444
•
111
```

If we add together a bond-pair's 4 and 1, their sum is 5. Ditto for all three bond-pairs, so this 6-pack adds up to 15. As a cross-check, we can add the same numbers another way...this time, by reading across the 6-pack. Its top half is 4 + 4 + 4 = 12. Its bottom half is 1 + 1 + 1 = 3. Now add the two sums together: 12 + 3 = 15.

$$4+4+4=12$$
$$1+1+1=3$$
$$5+5+5=15$$

In fact, this sum of 15 exists for each 6-pack on that bottom row of the chart. All along its row, every 6-pack adds up to 15!

444	442	424	422	244	242	224	222
111	113	131	133	311	313	331	333
15	15	15	15	15	15	15	15

Not only that, across the whole chart, any 6-pack relates and balances with every other 6-pack. If we add up a bonded pair of numbers anywhere on the chart, such as 4 • 1 or 3 • 2, its total is always 5, and its 6-pack always adds up to 15. In other words, this chart shows 64 different ways of saying 15! Why? It happens because the balancing relationships inside the chart are correlated with hidden binary aspects of the underlying co-chaos paradigm.

All 64 discrete 6-packs of additive numbers separately rush to summary judgments, but they all reach the same conclusion: 15. Thus DNA's double helix is full of magic squares of 15! Shades of the ancient Luo Shu from China in 2200 BCE! Its numbers in a "map" form a magic square of 15. Here our magic square of 15 also appears in the 64 molecular 6-packs of DNA, hidden in life itself! (See Chapter 11 for more discussion of the Luo Shu's magic square.)

What appears to be just additive summation somehow also has many proportional relationships going on inside it. How very rhythmic is its number distribution. For instance, each bond-pair in a 6-pack may be viewed as a ratio. Thus the bond-pair of 4 •1= 4/1. The bond-pair of 2 •3= 2/3. The bond-pair of 1• 4= 1/4. If you follow this out across the chart, the result puts consistent ripples into its number fabric. Tricky, this shifting emphasis between linear, goal-driven sums and relational, process-oriented ratios. Both modes are at work to hold together this chart.

As I view the number chart, I find various events playing across it. Wherever I look, I see goal-oriented aspects and process-oriented aspects doing their complementary dance. It opens deepening perspectives in a series of windows.

This system's structure is so complex, yet so simple, that I cannot reach all its implications. I'm not going to show you any more of the intricate simplicities I found inside this number chart, but you can explore its echoing network as far as you care to go.

6. DNA Bolyai Quilt

Now we come to the *DNA Bolyai Quilt.* Its symbols (and in the ebook version, also its colors) represent the 64 hexagrams of the I Ching, or equally, 64 molecular 6-packs in a double helix. It uses the same binary organization as the last chart, but we are now considering the design patterns rather than number patterns. Here each 6-pack is a pair of 3-packs that bond via the bindu symbol of ⊗. Now the 64 hexagrams—or the two bonded strands of parental DNA—appear on this chart as a succession of quilt blocks.

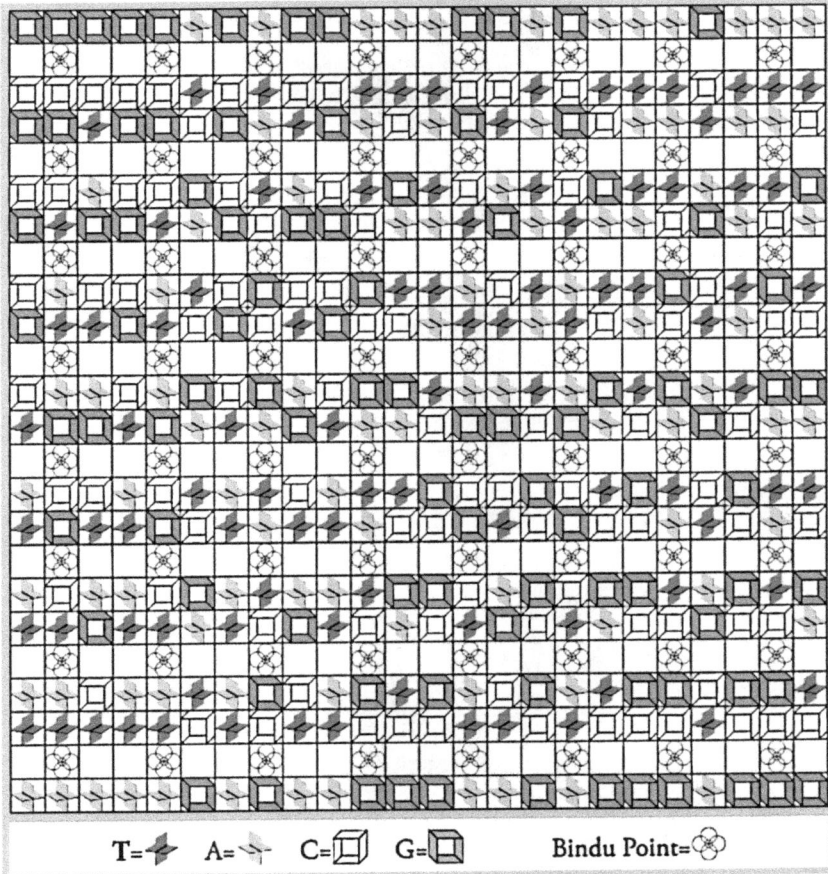

DNA Bolyai Quilt & code key

7. Chinese Silk Tapestry

This next design is also based on the same DNA chart, but by choosing new symbols and colors, the effect becomes quite different. It suggests to me a Chinese tapestry in brilliant silk. It has the sumptuous quality of robes worn in a Chinese opera that I saw in Guangzhou in 1991. I was fascinated by the gorgeous garments, stylized makeup, gestures, eerie sounds, and all those carry-in picnics that the audience members were enjoying.

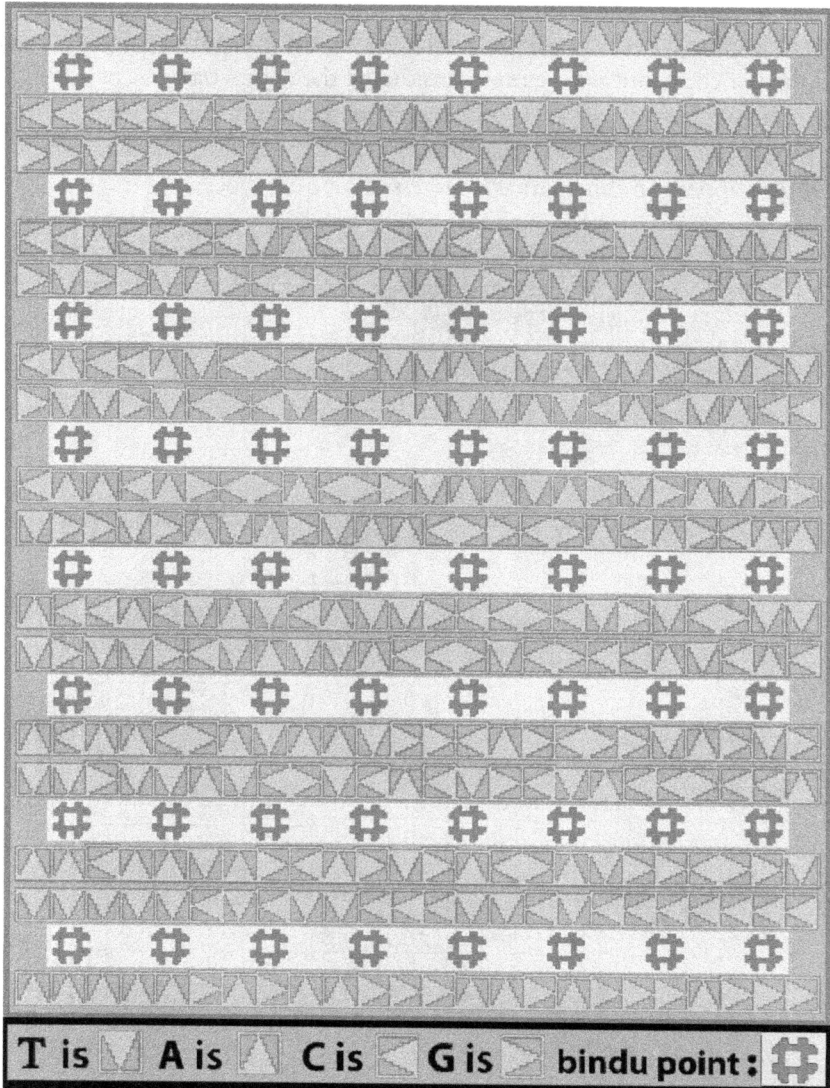

Chinese Silk Tapestry & code key

8. Good Seas

Now I'll show you the same organizational chart yet again, and moreover, using the same symbols, but now the design is re-colored in a way that changes the result. I like this rendition for its clean crispness.

I can imagine this design as a bedspread woven in white, blue, and green. I call it *Good Seas* because something here suggests pennants rippling their semaphored messages across the ocean waves. It reminds me of when I sometimes looked out from the harbor at ships passing in the South China Sea.

Good Seas

This green bindu point 🔵 bonds the two codons in each DNA 6-pack, and it bonds the two trigrams in each hexagram. A modern Westerner might call this symbol of ⊞ a hashtag #, a pound sign, or a tic-tac-toe diagram. But in ancient China, this very old symbol on a map indicated a community around a well, so it became the Chinese symbol for Hexagram 48 *The Well.*

In ancient China, the governing decree divided a parcel of land by cross-hatching it so that eight families could live on the eight plots of land surrounding a central ninth portion kept for public use and shared in common.

That central portion of land usually held the communal well. It was owned by all together, and people gathered around the well to quench their physical and social thirsts. Hexagram 48 *The Well* symbolizes the concept of endless renewal from the depths, of drawing up revitalizing buckets of community strength from the water table of collective identity that is hidden below ground-level events. This hashtag method of land division gave all equal access to the well in common; it was in effect a social rendition of the magic square.

9. Laughing Gas Afghan

I call this next design the *Laughing Gas Afghan.* Why? It was only after I'd finished making it that I finally realized this design would give the effect of so many laughing, gabby mouths. What a gas! Looking at it made me grin in a giddy way to imagine all those open mouths yukking it up.

My grandmother, Jewell Wilson, knitted many afghans. She might have called this design *Gossip Girl.* She was born in the horse and buggy era. Her parents rode cross-country in a wagon to live in a sod-topped house that they dug into the prairie soil of the north Texas panhandle. Granny herself, whom I only saw as an old, crippled woman, was born there. When she married my grandfather, Oliver, the neighbors helped them dig their own sod-topped house. They lived in that dugout house for about 12 years. Life was hard.

But then as a farmer's wife, Granny eventually lived in an actual lumber-built home with a wooden outhouse. In her spare time—you know, when she wasn't building a fire under the wash pot outdoors, canning, gardening, scouting the hen house for eggs, and chasing out snakes—she'd quilt, knit, or crochet in a sewing circle of friends. I have a Friendship quilt that they all hand-sewed as they sat gossiping together around the quilting frame. Its patchwork is embroidered with their names, including two of my relatives of a modern-day Texas governor.

Local gossip fueled their talk, and from my grandmother, I heard many tales from back in the days when she and her friends put their heads together over a quilt, afghan, or baby blanket they were making for some invalid, elder, or infant. She lived on into the time when radio, telephone, and TV turned daily gossip into something no longer local but national and even international.

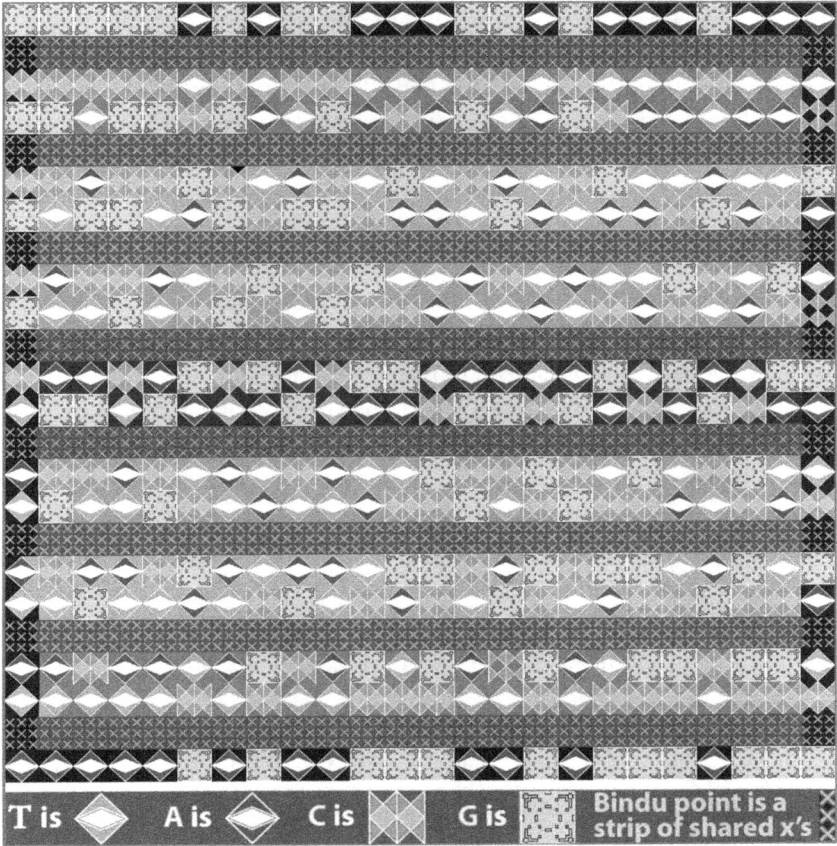

Laughing Gas Afghan & code key

10. The fabric of the universe

Yes, the smiling lips of this *Laughing Gas Afghan* remind me of Granny and her sewing circle. The continuous cross-knit of tiny x's representing bindu points seem to me like a connective web that bonds all those mouths into community. Or maybe each tiny x in the cross-knit weave suggests the kisses that keep us all sharing, working, playing, and living together. Nature's interwoven numbers are in us, around us, holding us in concert.

It is no wonder that Indonesian mythology said our universe is made of cosmic threads whose warp and woof weave the design of its great fabric. In fact, many older cultures employed a vocabulary of fabric-making with metaphors of carding, spinning, and weaving to talk about universal creation. Modern physics does it, too, in studying "spacetime fabric" and "string theory."

Members in my dream groups occasionally report having a dream where a great fabric spreads across the sky. It is a natural imagery that resides deeper than society's conscious connections. Its billowing gossamer net is a dream-time analogy for connectivity in the collective unconscious, the greater all, the Tao. This folk tapestry is made of all individual beings woven into the universal being. Such images in dreams hint at a unity that weaves everything together.

I had such a dream once, back in 1988. In it, I saw famous people whom I admired and somehow knew as friends. I saw that they were making wind sculptures out of filmy, colored fabrics. Those ballooning sculptures moved and fluttered across the sky like some huge version of Tibetan prayer flags.

Hmm, why did I have that dream of famous people busy at soul-making? Well, everyone I know is famous to me, and always busy at soul-making.

A dream is holographic, so after you wake up, even if you remember only a fragment of what occurred in it, examining just that fragment can reveal a fuzzy image of the central message it holds for you. The more details you can manage to remember, the more that message deepens, refines, clarifies.

This deepening trait is the essence of co-chaos patterning. It generates the flux of space, time, matter, and energy that we live in, that we are. The vast hologram that is our Double Bubble universe is real, and we are real... as real as anything inside this giant, shifting, evolving life form. Its grand organizing design is endlessly ideal, given the parameters that our universe has to work with. Wherever you look, co-chaos opens up windows onto new perspectives...especially if you realize the DNA double helix that Watson and Crick found in 1953 is based on co-chaos math recorded 3 millennia ago in Chinese hexagrams.

Such synchronicity is cued by webby connections that mostly billow at the unconscious edge of life, signaling subliminal messages from the background fabric of events. In the tiny fraction of reality that is apparent to our senses and logic, we blithely conduct our lives in a way that takes this universal support system for granted. But that's okay. The universe is used to that, prepared for that. It loads a lot of redundancy into the system to deal with that.

We don our molecular bodies for this brief duration in space, riding on the arrow of time here in our upper bubble. At night, dreams help us tap into complex patterns stored in the great unified mind in the lower bubble's 3D timing. We're all participants in this design. We sew our individual textures into the larger whole, inserting odd bits of lace and tangle, leather, sequins, or rickrack here and there. Our filmy souls are busy making wind sculptures that fly among the others. All of this weaves each unique life into the cosmic fabric.

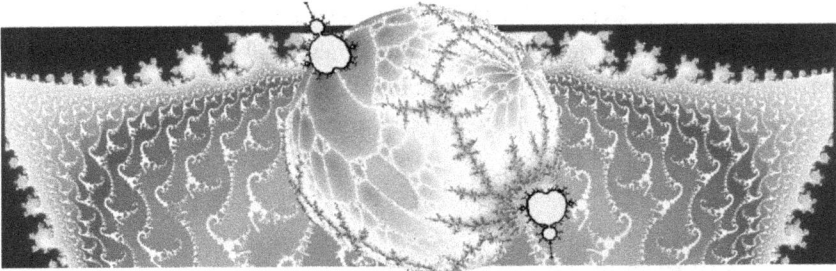

Chapter 9: RNA as Design

1. Beyond DNA to RNA

You recall the RNA codons in the recursion puzzle from Chapter 7?

CCC	CCU	CUC	CUU	UCC	UCU	UUC	UUU
CCA	CCG	CUA	CUG	UCA	UCG	UUA	UUG
CAC	CAU	CGC	CGU	UAC	UAU	UGC	UGU
CAA	CAG	CGA	CGG	UAA	UAG	UGA	UGG
ACC	ACU	AUC	AUU	GCC	GCU	GUC	GUU
ACA	ACG	AUA	AUG	GCA	GCG	GUA	GUG
AAC	AAU	AGC	AGU	GAC	GAU	GGC	GGU
AAA	AAG	AGA	AGG	GAA	GAG	GGA	GGG

RNA Codons in binary order

The same codons appear below, but now they're written in the glyphs used in Chapter 8 for DNA 6-packs in *DNA Bolyai Quilt*. However, since Chapter 9 focuses on RNA, now the symbols below depict RNA 3-packs. Each T changes to U, and its title shrinks to *RNA Bolyai Throw*.

Or hey! This image below can also symbolize the lower trigrams in Shao Yong's binary order of hexagrams.

RNA Bolyai Throw & code key

2. Heard it Through the Grapevine

Here is the same codon order again, but in new symbols. As this pattern was emerging on my computer screen back in 1989, a tune popped into my head.

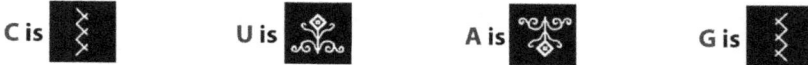

Heard it Through the Grapevine & code key

I realized its viney sinuosity was provoking Marvin Gaye to sing between my ears: "Heard It Through the Grapevine." This coding variant apparently stimulated networks of association that went musical in my brain.

The brain often responds to the genetic patterns embedded in nature in ways that go more sensory than sensible, more emotional than logical, more arty than smarty. Indeed, perhaps most of us access the rhythms of nature mostly with our sensory or artistic appreciation rather than by logical thinking. We just vibe to it as we're running in the park or playing with the dog.

3. A Visual Symphony

The next image uses the same codon order, but again to different effect. Unfortunately, black-and-white print cannot convey the transparent symbols floating on a black surface to reveal a prismatic color bed that vivifies the ebook version. Since black print turns the dots of color into speckled white lines, nullifying their lively effect, it seemed reasonable in the print version to instead make the image's white lines more clear and absolute.

A boomerang shape is the only symbol used in this image. Its wedges can sit closer together or farther apart, and they can also mirror-flip to either side.

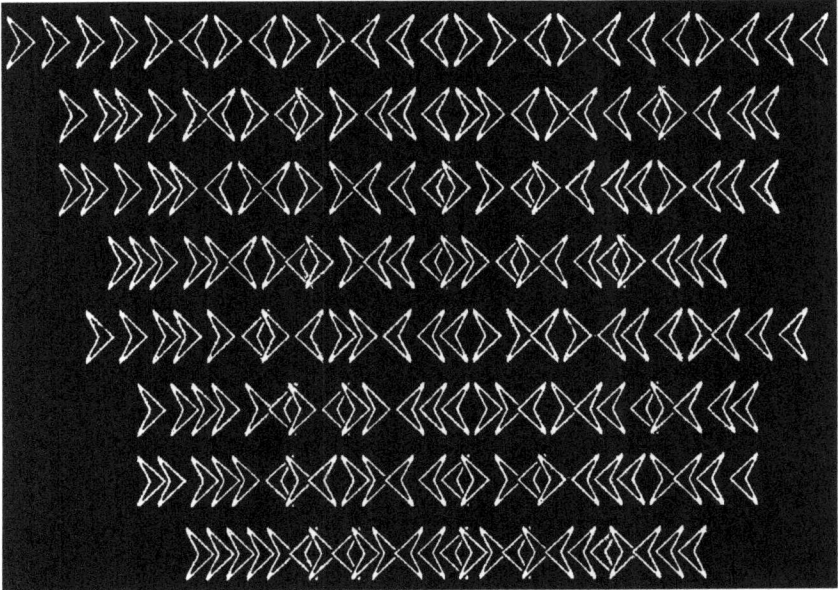

Visual Symphony

Variable spacing between the boomerangs, along with their mirror-like flip-flops, give us the only means to differentiate the four different kinds of molecules. So here's a tip: start at the top left and move counterclockwise around the image. At each corner sits one of the four mono-codons: CCC, AAA, GGG, or UUU. They show the most uniform expression of each wedge's position and spacing, and they correlate with the *Basic Symbols* column below.

The second column, *Pairings*, shows how different molecules will sit by each other in a codon. The wedges may overlap in the design, or they may distance each other. For instance, C sits by G as adjoining, reverse-mirrored opposites.

Basic symbols	Pairings	Traffic Codons
CCC =	C by G =	AUG =
AAA =	G by C =	UAA =
GGG =	A by U =	UAG =
UUU =	U by A =	UGA =

Visual Symphony code key

However, when A sits by U, those same reverse-mirrored symbols will quit adjoining and instead exaggerate the distance between that earlier C-by-G pair.

The third column, *Traffic Codons,* shows the four traffic codons of AUG, UAA, UAG, and UGA; these four tell a ribosome when to start or stop working on the RNA strand. Since a trio of molecules forms each traffic codon, it is coded by a trio of wedges. By becoming alert to the spacing variables in the four traffic codons, perhaps you can spot and read the other codons, too.

The wedges seem to impart a sense of motion because of their fluid spacing, and as I was developing this design, that movement began telling me a story. The funneling wedges tightened as my eyes moved down the graphic, and the progressive contraction of its hieroglyphics gave me a sense of tension gathering to a climax at the bottom. It reminded me somehow of a symphony with people sitting in the audience, listening at first with their hands folded in graceful attitudes of pleasure, and then at the end, they start clapping.

On that gray Sunday afternoon in Kusnacht, a suburb of Zurich, that's what came to mind as it emerged on my computer screen, so I decided to call this image *Visual Symphony.* Then I had a wild idea of adding some colors by turning the white symbols transparent over a black background and then hiding a rainbow wash beneath. But in 1988, my printer only printed black. So why bother? I quit work and took a break to walk the kinks out of my back.

Twilight was coming. I wandered around Kusnacht, looking on a whim for some diversion. Oops, suddenly I realized that I'd left my wallet at home. Okay, I couldn't do anything that required money. Then on a corner wall, I saw a sign posted for a free concert featuring harpsichord, flutes, and other period instruments. I realized it was only two blocks away, and it started in five minutes!

In the light snow, I hurried over to a nearby Catholic church and was right on time. I entered a pew and sat down in an audience of maybe 80 people. The music of Farnaby, Vivaldi, Bach, and Bartok filled the air as I imagined graceful, flying boomerangs of sound cavorting in the high, dark space above us, with their washes of notes resounding like rainbows off the stone walls.

Then in 2008, Adrian Frye showed me how to use Photoshop to turn the white lines of the wedges transparent on a black surface and then add a layer of random color wash beneath the black to let the colors show through from below, creating a result I'd envisioned almost 20 years before. It vivified the codon wedges, and somehow that matters to me…although the colors are not code signifiers and merely add a rainbow brightness to the boomerang shapes.

4. Swedish tiling…and a modern Amerindian rug
This next take on the RNA codon chart feels more light-hearted and playful.

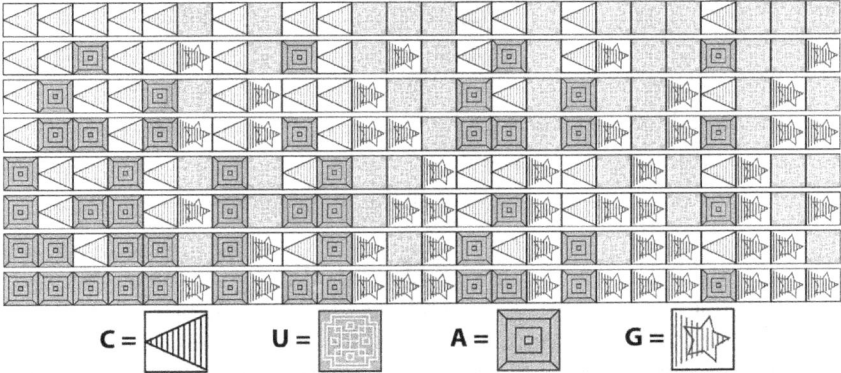

C = ◁   U = ▣   A = ▢   G = ✦

Swedish tiling & code key

This design has a clean, tidy, yet cheerful effect. It reminds me of a tiling frieze along a wall in Sweden. Look! The C and G symbols are both pennants, and G's pennant even holds a star! Those symbols could even go on children's blocks.

Do you have a preference?…that lighter, more whimsical tiling above… or below, the tighter, bolder, more rug-like design?

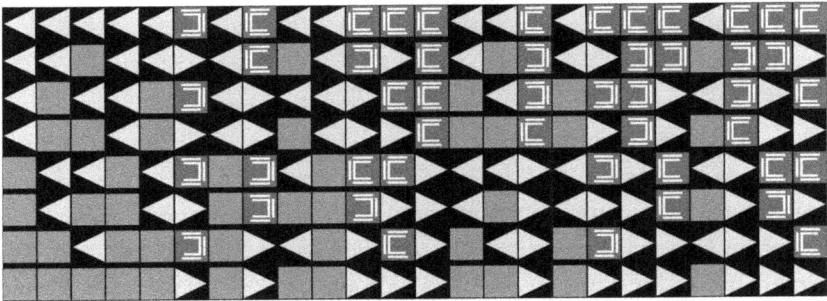

C=◀   U=⊏ or ⊐   A=▮   G=▶

Amerindian Rug & code key

Both designs employ basically the same shapes, but there's far less detail in the rug version, making its effect more emphatic. Part of that bold, blocky impact came from a printer glitch. In 1988, my HP printer balked at printing out the original square symbol of ▣ that I chose to represent U. Instead, it would randomly print out two partial versions of a square—either ⊐ or ⊏.

I was irritated at first. Then I realized the glitch was fortuitous; it offered me a real U to symbolize U…just tilted to either side! So I went along with the quirk and let both versions of U stand for the Uracil molecule.

This rug's bold pattern suggests to me a modern take on a traditional Amerindian rug. Hmm, if I repeated that rug four times…and then sewed all four rugs together in a reversing mirror-image...they'd join at the center to create a new, larger, fully symmetrical rug.

Folk rugs were often made in just that fashion. In a previous rug-making incarnation, I'd have matched my rug blocks at the lower left-hand corner, which would put all the solid blocks in the center and run most of those sidewise U's out toward the edges. But in this lifetime, I'd just go to a rug store.

5. Mirroring Quadrants

Matching those four quadrants together to form a larger rug reminds me of something that occurred with the I Ching when I was living in Switzerland. One day, a physics graduate student at the Eidgenossische Technische Hochschule stopped me just as we were both about to pass through a doorway. Why? He'd noticed that I was holding an I Ching book.

He stared at the book in my hand, pointed, and said he'd seen that same book cover on display in a shop window near the university. It was a paperback of the James Legge translation. He said the book cover stopped him cold on the sidewalk outside the bookstore, said that he'd stared at it for some time, trying to figure out those strange symbols inside the circle on its front.

It was hexagrams that he was staring at, of course, attracted by the obscure mathematical symmetry of their solid and broken lines. With an odd fervor in his eyes, the student went on to tell me that he'd known nothing of the I Ching back then, yet he nearly went in and bought that book. Just for its cover.

His name was Urban Studer, and the impact of those polarized lines marching in permutation across the gold and red background had a wordless power on him that was provocative and mystifying. He said he'd begun to explore the I Ching and found in it a profound depth of philosophy behind the simple yin and yang symbols.

After that encounter in the doorway, I finally examined the cover of my own Legge book more closely. Yes, okay, I admit it. For years, I'd just taken that cove's image for granted as...well, some hexagrams on a book cover.

But now, a closer look counts out 36 hexagrams. All different? No, those 18 hexagrams on the left are replicated in mirror-image by the 18 hexagrams on the right...as if a vertical mirror is reflecting the same 18 hexagrams twice.

Or no, wait a minute! Maybe...does a horizontal mirror reverse the 18 hexagrams above into the 18 hexagrams sitting below?

Or wait!...can it instead be that all 4 quadrants hold the same 9 hexagrams, with them all arranged as vertical and horizontal reversing mirror images of each other?

Or...can I also see this design as two diagonal flip-flops that mirror-reverse

the same 9 hexagrams into 4 four different quadrants?
Or all of the above? Kudos to the cover maker!

THE I CHING
THE BOOK OF CHANGES

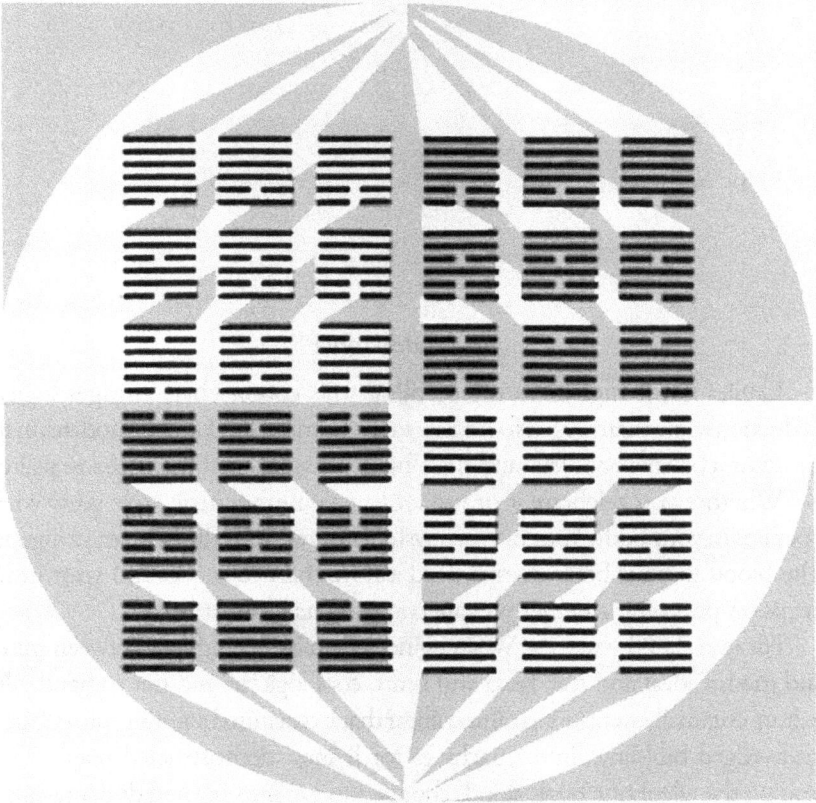

TRANSLATED BY JAMES LEGGE

1963 cover of Legge translation, Dover Books

6. Codons in a biology text forgo art

If you took biology, you saw an RNA chart like this one from Wikipedia:

1st base	2nd base				3rd base
	U	**C**	**A**	**G**	
U	UUU (Phe/F)	UCU	UAU (Tyr/Y) Tyrosine	UGU (Cys/C) Cysteine	U
	UUC Phenylalanine	UCC	UAC	UGC	C
	UUA	UCA (Ser/S) Serine	UAA Stop (*Ochre*)	UGA Stop (*Opal*)	A
	UUG	UCG	UAG Stop (*Amber*)	UGG (Trp/W) Tryptophan	G
C	CUU (Leu/L) Leucine	CCU	CAU (His/H) Histidine	CGU	U
	CUC	CCC	CAC	CGC	C
	CUA	CCA (Pro/P) Proline	CAA (Gln/Q) Glutamine	CGA (Arg/R) Arginine	A
	CUG	CCG	CAG	CGG	G
A	AUU	ACU	AAU (Asn/N) Asparagine	AGU (Ser/S) Serine	U
	AUC (Ile/I) Isoleucine	ACC (Thr/T) Threonine	AAC	AGC	C
	AUA	ACA	AAA (Lys/K) Lysine	AGA (Arg/R) Arginine	A
	AUG (Met/M) Methionine	ACG	AAG	AGG	G
G	GUU	GCU	GAU (Asp/D) Aspartic acid	GGU	U
	GUC (Val/V) Valine	GCC (Ala/A) Alanine	GAC	GGC (Gly/G) Glycine	C
	GUA	GCA	GAA (Glu/E) Glutamic acid	GGA	A
	GUG	GCG	GAG	GGG	G

RNA codon table

Unlike artful rugs or music or tiling, this codon chart engages logical deduction more than analog feelings...so the human mind accommodates in its constant rebalancing of left- and right-brain attention that pays heed to reality.

Whether you celebrate it or not, your analinear genetic code grew what operates in your body. Your nervous system buzzes. Your digestive tract digests. The blood in your body doesn't need a flow chart to move, and your heart keeps on pumping whether you are counting its beats or not.

For over 2,000 years, the West engineered an increasing split between mind and matter, ideal and real, head and heart, cool logician and hot romantic. Its culture eventually went so logical and linear that it denatured painting into cubism, knife-edged buildings into "machines for living," deconstructed poetry into reading the telephone book aloud, chopped music into isolated discontinuities of raw noise...as meanwhile, philosophers declared life to be meaningless.

But now humanity is seeking a wider culture that unites new and old, high and low, left and right, East and West, objective and subjective, linear and analog, head and heart, ideal and real. We seek a path that will give cool romantics and passionate logicians a transcendent third way into a much-needed update on our sciences, philosophies, aesthetics, and religions.

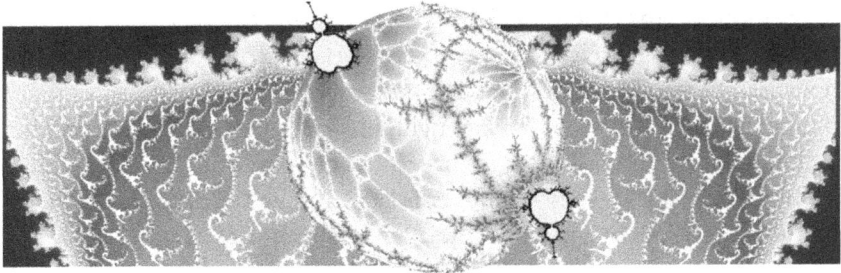

Chapter 10: RNA Task Hexagrams as Designs

1. Number Patterns in the RNA task order of hexagrams

By now, we've turned DNA/hexagram 6-packs into sheaves, a quilt, and an afghan. We've turned RNA/trigram 3-packs into grapevine, tiling, and a rug.

In this chapter, we'll turn the RNA task hexagram chart of the 20 amino acid families into genetic code trees. Why? The process will help us spot some interesting number patterns highlighting some key aspects of the co-chaos paradigm's inherent balance.

Ancient China used nature's traits to suggest the dynamics of the 8 trigrams, as shown on the chart below. It correlates the 8 trigrams in Shao Yong's binary order with paint's reflective color wheel, as discussed in Volume 2, *Co-Chaos Patterns*. (King Wen's analog order correlates with the projective color wheel of light).

Black Earth	Purple Mountain	Blue Water	Green Wood-Wind	◄ Paint's reflective color wheel
				◄ Chinese nature dynamics
				◄ Trigrams in binary sequence
0	**1**	**2**	**3**	◄ Digital equivalent
Yellow Thunder	Orange Fire	Red Lake	White Heaven	◄ Paint's reflective color wheel
				◄ Chinese nature dynamics
				◄ Trigrams in binary sequence
4	**5**	**6**	**7**	◄ Digital equivalent

Binary trigrams & digital equivalents

The next chart correlates the 8 trigrams in binary order with digits, colors, Chinese nature traits, and symbols in the first design, *Herald Tree*.

0 Black Earth	1 Purple Mountain	2 Blue Water
3 Green Wood	4 Yellow Thunder	5 Orange Fire
6 Red Lake	7 White Heaven	Bindu point

Herald Tree Code Key

Below left is the chart of 64 RNA task hexagrams, but with some updates. Each binary hexagram is now topped with its digital equivalent, showing you how this cross-coding into RNA task hexagrams has rearranged Shao Yong's binary sequence. The Column Key, right, holds what looks like fractions, but each numerator and denominator have a bindu point instead of a line. Each number stands for a binary trigram translated into its digital equivalent. Each bindu point bonds two trigrams into a hexagram. It's a *bindagram*.

The top set of 8 bindagrams is $\frac{20202020}{55441100}$. Just two numbers alternate above the bindus: 20202020…this keeps adding or subtracting by 2. But below the bindus, numbers count down by pairs: 55441100…and skip two middle pairs.

Column Key — Bindagrams (trigrams are digits):

RNA TASK ORDER — 64 RNA Task Hexagrams & Binary Equivalent	Bindagrams
PROLINE 42 40 / LEUCINE 34 32 / SERINE 10 8 / PHENYLALANINE 2 0	20202020 / 55441100
43 41 / 35 33 / 11 9 / LEUCINE 3 1	31313131 / 55441100
HISTIDINE 46 44 / ARGININE 38 36 / TYROSINE 14 12 / CYSTEINE 6 4	64646464 / 55441100
GLUTAMINE 47 45 / 39 37 / 15 STOPS Ochre, 13 STOPS Amber, 7 STOPS Opal / TRYPTOPHAN 5	75757575 / 55441100
THREONINE 58 56 / ISOLEUCINE 50 48 / ALANINE 26 24 / VALINE 18 16	20202020 / 77663322
59 57 / 51 49 START / 27 25 / 19 17	31313131 / 77663322
ASPARAGINE 62 60 / SERINE 54 52 / ASPARTIC ACID 30 28 / GLYCINE 22 20	64646464 / 77663322
LYSINE 63 61 / ARGININE 55 53 / GLUTAMIC ACID 31 29 / 23 21	75757575 / 77663322

RNA task hexagrams–binary trigrams–column of digital equivalents

2. Herald Tree

The Column Key of *bindagrams* translates into the branches of *Herald Tree*.

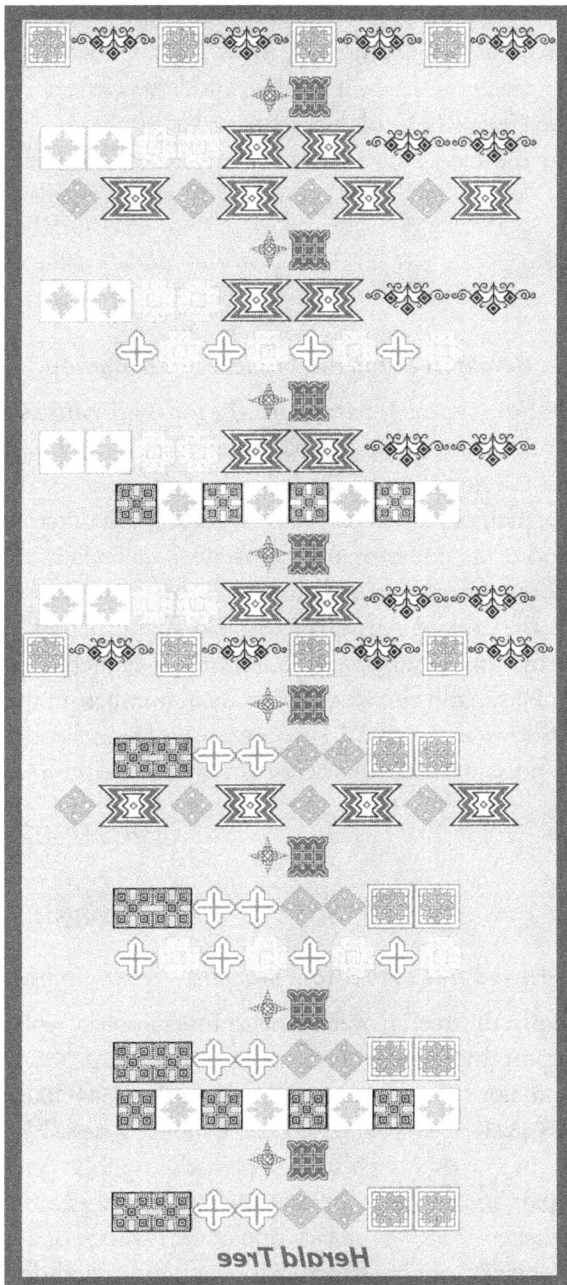

Herald Tree

3. Some Branches of the Herald Tree

This genetic code tree is named *Herald Tree* because its symbols remind me of heraldry, or old playing cards, or maybe those alchemical symbols in an esoteric, medieval book. Although *Herald Tree* is rendered in a symbolism reminiscent of medieval times, it catalogs modern science's 20 amino acid families grouped into RNA task hexagram order.

Throughout the tree, a 2-ness plays in binary, analog, and analinear rhythms.

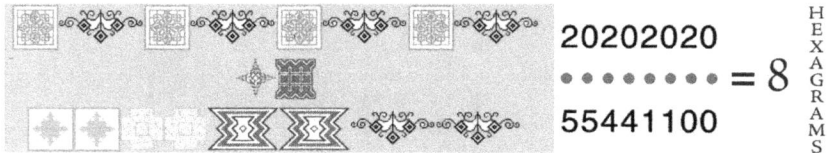

20202020
•••••••• = 8
55441100

H E X A G R A M S

Herald Tree's top two branches = 16 trigrams

Each symbol/digit stands for a trigram. The top two branches in *Herald Tree* hold 16 trigrams that bond into 8 hexagrams. The sole bindu point bonding all 8 hexagrams is a single star beside a chunk of tree trunk .

The over-branch or *numerator row* is 20202020, thus alternating just two numbers: 2 and 0. Its under-branch or *denominator row* is 5544 (skip 3, skip 2) 1100, thus decreasing by paired numbers…and also skipping two pairs of numbers in the middle. Across the whole tree, this pattern persists. Each over-branch's rhythm *alternates* by 2s, but its under-branch's rhythm *decreases* by paired numbers…and also skips two pairs of numbers in the middle.

Moving on down the tree, each successive pair of branches has 16 trigrams. They're pair-bonded into a row of 8 hexagrams by its star-and-tree-chunk hunk.

20202020
•••••••• = 8
77663322

H E X A G R A M S

Herald Tree's 9th & 10th branches = 16 trigrams

Halfway down the tree, an over-branch/numerator row replicates that first over-branch uptop by echoing the same sequence: 20202020. But its under-branch/denominator row differs: 77663322 versus that 55441100 uptop.

Hmm, 77663322 - 55441100 = 22222222. Echoing 2-ness. Now I'll bracket their numbers. In 77663322, each bracket-pair adds up to 9. But in 55441100, each bracket-pair adds up to 5. Okay, 9-5=4…as in 2+2…or 2×2…or 2^2.

This entire genetic code tree, in fact, ripples with a breeze of 2-ness, signaling the underlying binary number structure, but also the period-doubling typical of standard chaos theory. Ah, the 2-ness of it all…a recurring motif in co-chaos.

4. City Tree

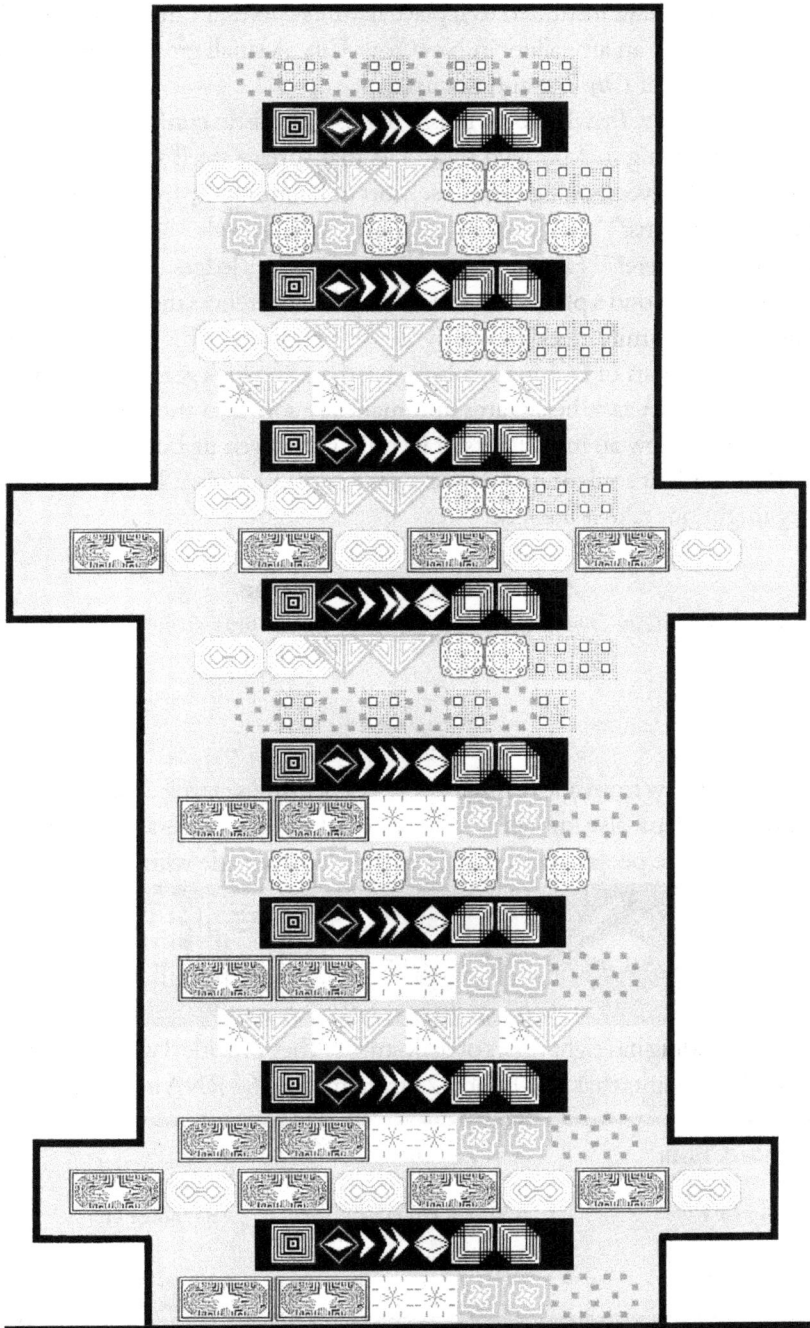

City Tree

This next design, *City Tree,* grows in the squared-off stone and asphalt jungle that we have instituted to replace nature's curves. I can imagine people walking around an art gallery in New York City. A small group stop before a graphic poster of *City Tree* and discuss it.

"Hmm, *City Tree.* I like its brash, almost tribalistic confrontality." "Did I see this building in Midtown East?" "Is that surface steel? Or aluminum?" "Hey, spot the eyes looking out at us…sort of haunting the ledges?" "Oh! Is it the eyes of golems?" "No, looks to me more like sand owls." "But what would they be doing here?" "Hey, hawks can live on the city's ledges. Why not owls?"

One reads aloud a placard that says, "*City Tree* references the modern urban jungle full of family trees grown by the genetic code. This image's skeletal structure is a chart of I Ching symbols shorthanding life's genetic code. Its 64 full-blown RNA task hexagrams automatically sort into the 20 amino acid families that grow all inhabitants of its building…even its bugs and rats."

People might smile and nod wisely, yet be none the wiser. Being so literate, they might not take it literally.

5. Branches of the City Tree

Look atop *City Tree* to find this sequence in the gray stone…

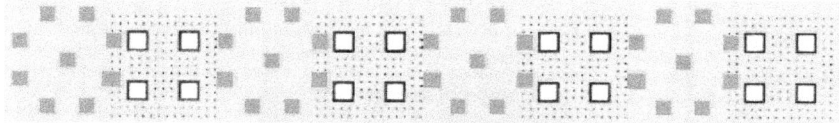

The sky view from above might see it as a penthouse terrace that alternates trees and gardens, also known as 20202020. Below that level is a floor of tinted windows where people stand talking as they look outside with binoculars.

The building has eight floors of dark stripes. They provide 8 windows. These act as bindu points that pair-bond 128 trigrams into 64 RNA task hexagrams.

0 Black Earth	**1 Purple Mountain**	**2 Blue Water**
3 Green Wood	**4 Yellow Thunder**	**5 Orange Fire**
6 Red Lake **7 White Heaven**	**Bindu point**	

City Tree Branches & code Key

The building's first four floors have window boxes bursting with a variety of flowers planted in pairs. In the Column Key, they are known as 66 and 33 and 22. Or we could call them the lower trigrams of ☰ ☰ and ☰ ☰ and ☷ ☷ inside the RNA task hexagrams. (This example can help us see how to decode all the following trees.)

 and and

But even in the city, the wild still prevails in places. Various plant and animal archetypes haunt the human collective unconscious and spill over onto the building's ledges.

Sometimes they even overlap, so when I'm looking at the ledges, I can imagine that ancient Aztec jaguars lurk on them behind the flower petals.

Or is that blood dripping from their jaguar jaws? Startled, I at first don't realize that the images symbolize four alternations of the trigrams of ☱ Red Lake and ☲ Orange Fire, nor do I interpret them as 65656565.

On the ledges of gray stone, some may see golems in the cubbyholes, but others find tawny sand owls nesting there. Owl imagery often emerges into human consciousness as a projection of totemic wisdom and insight peering at us from somewhere within the tree of life. In it, the symbol represents the trigram ☳ of Yellow Thunder (and Lightning).

In the image below, its bottom layer shows three owls nesting in cubbyholes between starry chunks of Heaven. A fourth owl sits at the right end of the ledge to maintain lookout. Above them, two more owls stand guard, but…

…oh no, it looks like one of the guard owls is getting chomped by a jaguar! Or is that a Martian invader?

That dark window stripe of the apartment above them shows an occupant looking outside with binoculars, yet never noticing what goes on just below.

6. Basketry
I like the folksy, East-European flavor of this next design. I call it *Basketry*.

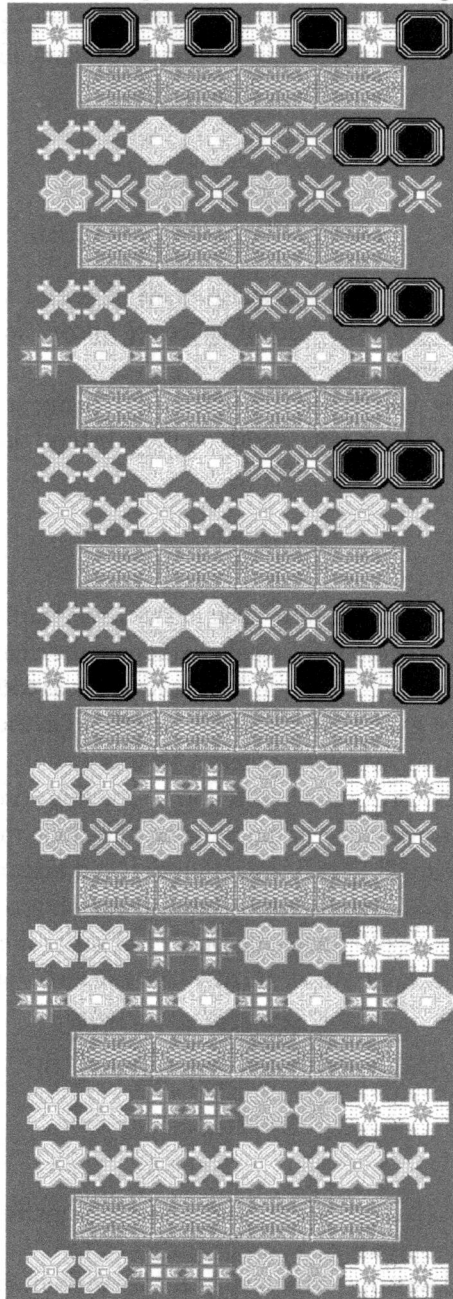

Basketry & code key

0 Black Earth	1 Purple Mountain	2 Blue Water
3 Green Wood	4 Yellow Thunder	5 Orange Fire
	6 Red Lake	7 White Heaven
	Bindu point	

Basketry is held together by 8 tiers of brown basketweave, filled with trinkets that symbolize the 64 hexagrams in RNA task order. Each tier of weave represents the bindu points that bond together its row of 8 hexagrams. The weave itself was a gift of synchronicity that came from hitting the wrong key on my computer. Oops! Hmm…hey, okay!

Let's select a tier of basket weave to examine its designs. Count down from *Basketry's* top to find the fourth tier of basket weave. It acts *en toto* as the bindu point for all 8 hexagrams along that row.

On the far right is a hexagram made by the lower trigram of ⬛ or ☷ *Black Earth* bonded to the upper trigram of ✖ or *Orange Fire* ☲. Taken together, this pair of trigrams symbolize Hexagram 35 ䷢ *Easy Progress*. Its philosophical directive fits well with Tryptophan, the amino acid that helps people relax after a turkey dinner.

To me, these symbols of ✖ *Orange Fire* and ⬛ *Black Earth* suggest the sun's activating rays beaming over the earth's foursquare stability, a pleasant notion from ancient China that gives a visual cue for Hexagram 35's metaphor of the sun rising in its natural, easy passage above the sturdy, reliable earth.

However, I did not plan this specific pairing of designs for Hexagram 35. The cross-coding match just turned out that way. Frankly, I chose *Basketry's* symbols according to their width and height, for I wanted to get a tall, narrow graphic like a columnar tree. It was just a happy accident that the result snuggled ✖ and ⬛ together like trinkets in the tiers of basketry.

7. Nature's numbers as images

The Arabic digits (1, 2, 3, etc.) adopted by the West do not emphasize the analog, qualitative aspects of numbers. For example, 2 does not show the doubling quality of 2-ness nearly so well as the old Roman numeral of II does. So why did we switch to the Arabic version? It was far more efficient at adding, subtracting, multiplying, dividing—all tasks that deal with quantity.

But what about the relational qualities of numbers? Their qualitative aspects usually get stripped away in everyday number tasks, and even more so in science. Nevertheless, nature itself is stippled with subliminal numbers that have a qualitative impact on us. If you look at a tree, you do not see numbers airbrushed into its surface, only the cantilevered limbs and ballooning foliage. Number's analinear play is submerged into the texture of that tree and its dwellers, iterating fractal patterns that push the forming edge of life to keep it vital.

Nature resonates at a level so deep that it taps into something below the level of human ego identity, deepening into the collective unconscious. Some societies have long prized this relational, connective quality in nature, and some societies resonated with nature more readily than others. The linear-skewed West has long recognized the Orient's traditional sensitivity to imagery in nature. It is evident in their calligraphy, paintings, architecture, clever cooking tools, and subtle silken garments. Such attunement bespeaks more than mere good taste. It is an analog awareness of "what likes to go together."

We are fortunate that ancient China handed down to us an amazing vision that connects mathematical co-chaos with philosophical directives described in nature's imagery. Image as word, image as design, image as impact. The 64 hexagrams are 64 dynamical directives hidden in a succinct shorthand.

We owe gratitude to the many people symbolized by Fuxi, Nuwa, King Wen, and others...those who found this co-chaos code, those who shorthanded it into I Ching hexagrams, those who memorized the math and words that finally became text, those who passed it along to us thousands of years later.

The basis of universal structure comes from co-chaos numbers submerged in the timing and spacing of all events. This *ur*-structure, in which all other structures can be embedded, underlies all the sciences, arts, technologies, crafts...all shapes, weights, colors, sounds...even every snort and sigh.

Generally, however, an artist does not view numbers thus. To the artistic sensibility, numbers may seem cold, mechanical, or soulless. Yet numbers can be passionately organic...if it's the special kind of analinear numbers that generate the genetic code. Its numerical gift utilizes both process and product; it values both relational quality and definitive quantity; it triggers both subjective and objective views. It creates action and reaction, flow and conclusion.

8. Other Trees

Marquetry

Marquetry

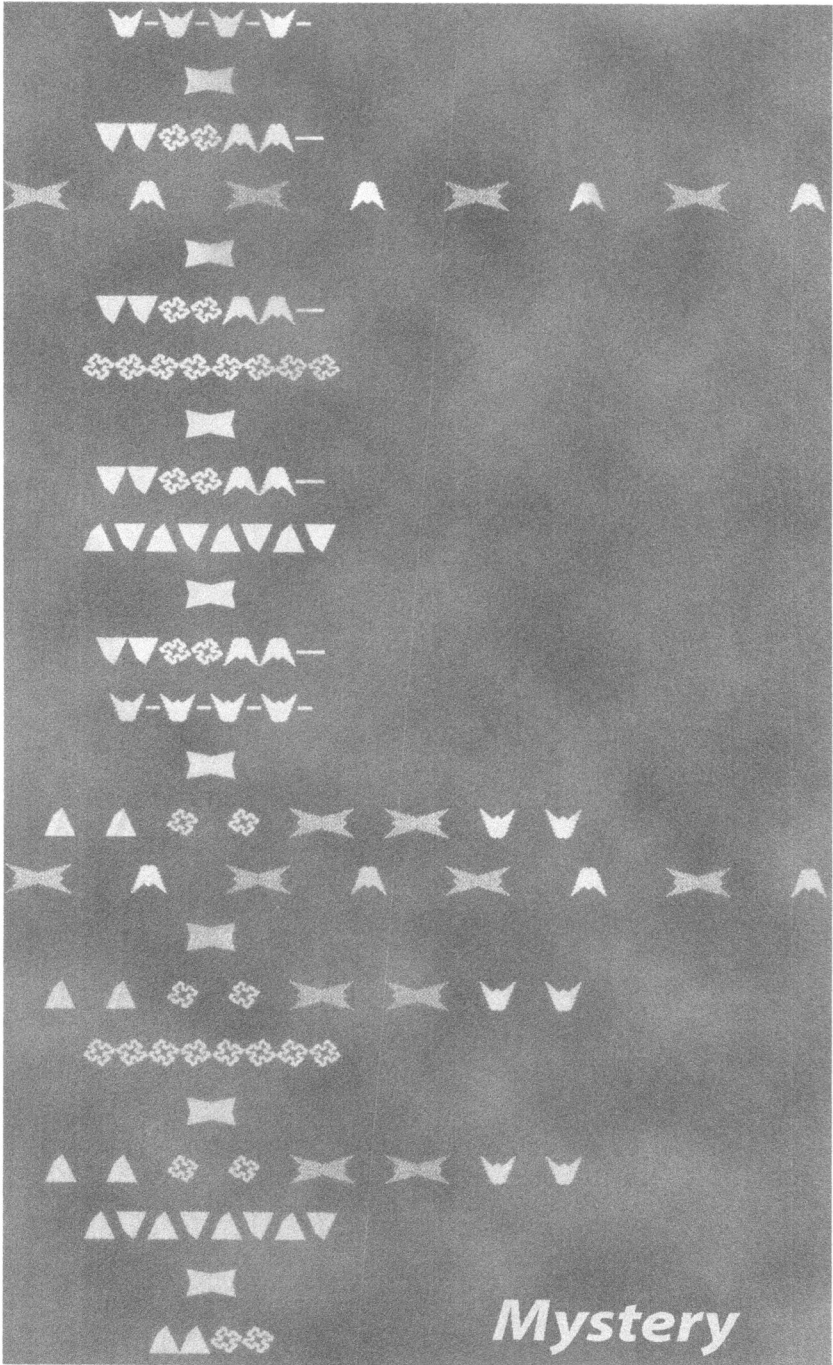

Mystery

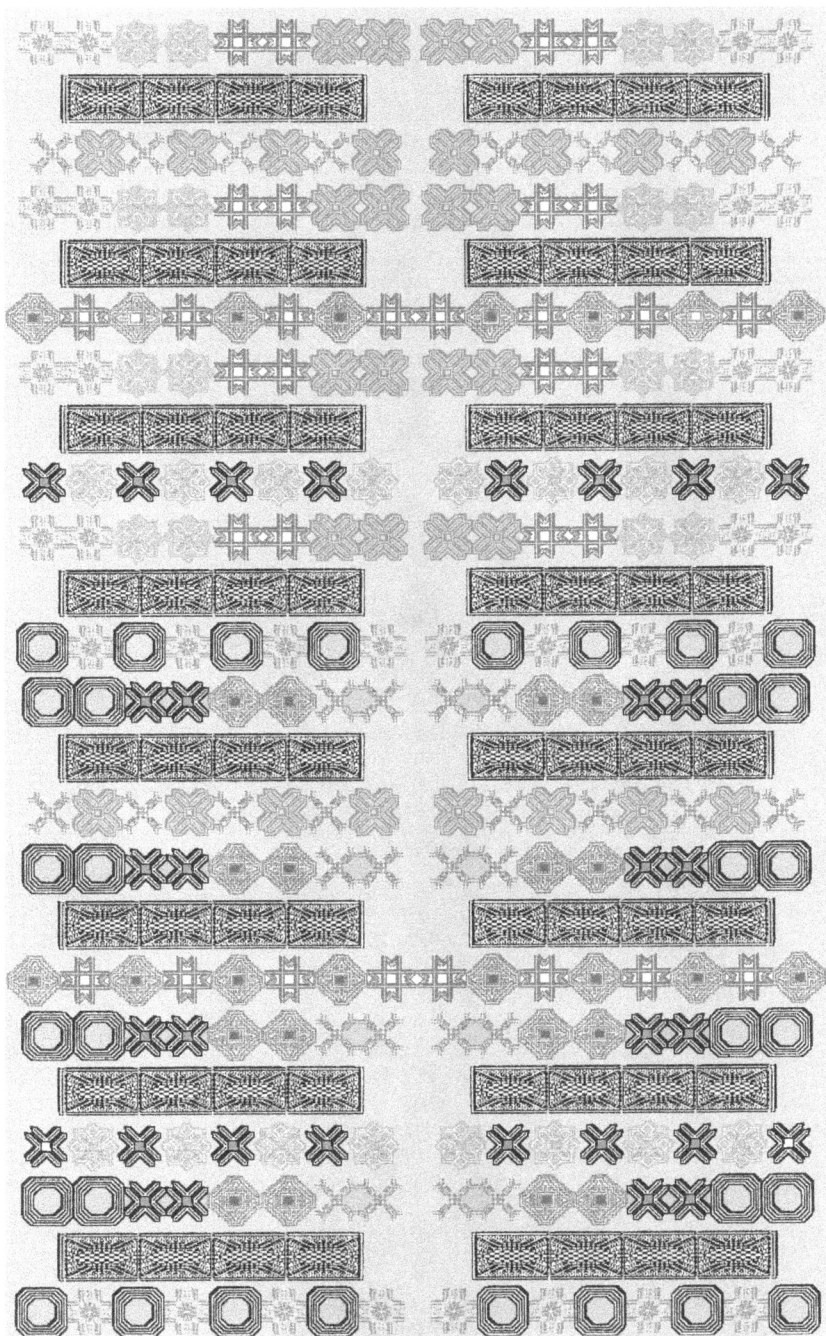

Mirror-Symmetry

ALPHA
CENTAURI

AISLE OF CEMETERY

LOFTY
SYMMETRY

ISLE OF
SUMATRA

I'LL HAVE SOME
TERRINE

I LOVE PASTRY

MY
STORY!

ENTRY

Mastery

For the last two graphics in this chapter, the 64 RNA hexagrams are embedded twice in each image, once above and once below, to create the design.

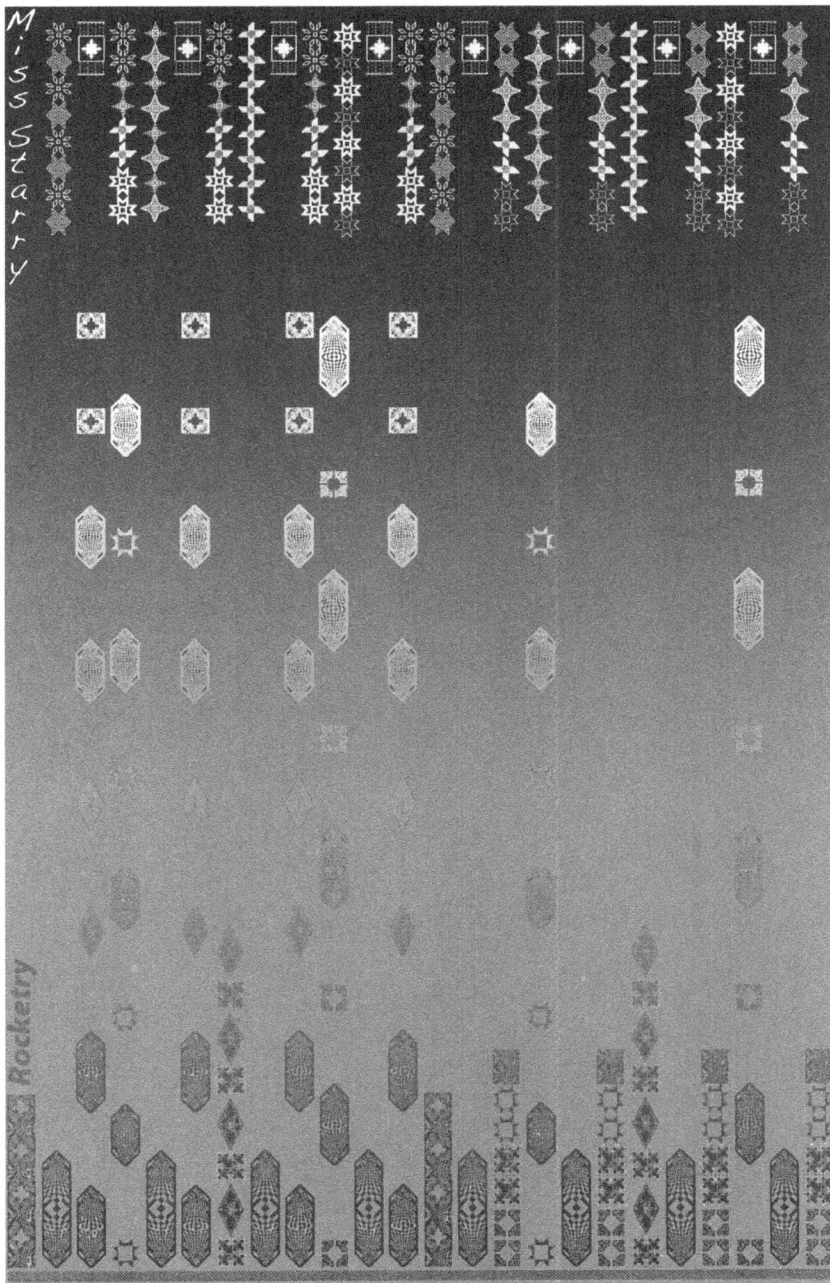

Miss Starry over Rocketry

Reflectory

Save a Tree over Reflectory

9. Nature is numbers in action

Those last 6 designs were included without comment to pique your imagination…or perhaps because it was just fun to make them. If you groaned at the oh-so-punny titles, don't forget that a pun, being the lowest form of humor, is also closest to the unconscious. The designs all code for co-chaos in the double p-tree of life, but they're expressed in various ways that emphasize shapes, colors, and movement more than numbers and biochemicals.

How did I manage to turn genetic code into the abstract designs of the previous three chapters? How could I come up with so many symbols? In Austin in 1988, I bought *Sunshine Font Collections 1 & 2* from a friend, Steve Schwartzman. He made the fonts and sold them on floppy disks at the University of Texas Mac Club. When I told Steve what I wanted to do with them, he gave me permission to use the fonts whenever, however, and wherever I choose.

As I sought some visual metaphors for how nature beautifies itself by putting hidden code into living cells, I simply typed some codon sequences on my computer in Helvetica font, then switched them into various fonts in the *Sunshine Font Collections*…which, alas, ceased to work when Apple changed how it handles type in 1991, so they became unavailable on the market.

Like these images, you are made of genetic code. But when you glance at your arm, or when you see someone walking a dog on a leash, or when you pass under a blossoming tree, you never notice the genetic coding level. Instead, you're only aware of an overall flow of muscle, motion, and mystery. You just think of these scenarios as nature at work. And it is at work. In numbers.

Often our current science chops up life and stacks it into the dead cordwood of linear data arranged in rows of discrete utility. But the webby cycling resonances in nature are what keep us alive, and its mysterious wholeness is what makes nature so hypnotically beautiful at every level of scale.

Take music, the most mathematical art. An octave of music is defined as "the unit of frequency level when the base of the logarithmic is 2." Cultures worldwide have developed different subdivisions of the musical octave in distinctive tonal scales of relationship to shape their music to various effects.

Jonathan Kramer points out in "Temporal Linearity and Nonlinearity in Music" that Western music is quite linear: "[In Western music…] The tonic is endowed with ultimate stability. All tonal relationships conspire toward one goal—the return of the tonic, finally victorious and no longer challenged by other keys. Thus tonal motion is always goal directed."

Goal-directed music has even created an international lingo in the media. Its subliminal aim: to ping your emotions. Now worldwide, music punctuates TV shows with the heroic swell, a drone of dread, some plucked heartstrings.

It glorifies the line-driving thrust of music in pursuit of its emotional goal. By now, car chase music in a Hollywood film parodies itself. So does that rush of notes pumping up the greed of a TV game show. I grew up on such music.

10. Western music, Turkish music, Chinese music

When I lived in Turkey in 1962-63, I found its culture was saturated with folk music that I could not even hear very well, much less understand, far less appreciate. In the little village of Yalova, I would ride on a bus and hear its sound system blaring music throughout its length, tuned to the local radio station. It rattled the whole bus with those heavy, stolid drum beats, shrill piercing of a *kaval* flute, a woman wailing in keening ululations that made no music to my unaccustomed ears. I thought it sounded peculiar and repellent.

Habituated from childhood to the heroic Western mode of music with its conflict, crisis, and resolution, I did not know how to surrender into this swirl of what seemed to me like repetitive, mournful cries, tediously prolonged. The music clustered in self-similar notes that cycled continually in hypnotic, not quite repeating patterns to my ears and mind. Such music does not really end. It reaches no final climax or resolution. Instead, it just trails off as if the pattern is sinking below conscious hearing into the brain.

At first, I could not let that endless process enfold me, not allow those cycling iterations to pull me into a flurry of emotions without any musical thrust of crisis and resolution...until suddenly the music began to intrigue me with its circling textures, envelop me in a whole new way, capture my feelings lost in the process rather than pointed toward a resolution. Finally, it entranced me by its poignancy of notes that were always just becoming, not just being.

That iterating yet evolving cascade of music apparently originated in a trance cult whose influence migrated westward into Greece, bringing with it the wanton, pleasure-loving god Dionysus. A wreath of iterating grape leaves encircled his brow. Dionysus was inebriated and ecstatic, body and soul, lost to ordinary logic. Maenads, his female followers, were the original groupies. Their music of flutes and drums enthralled listeners, triggering passions, taking followers into a deep swoon or mad revelry or exquisite sorrow.

Dionysus was seen as the dramatic opposite of law-giving, order-loving Apollo, the dependable sun god who daily drove his gleaming chariot across the sky. (Their polarity appears in more detail in Volume 2, *Co-Chaos Patterns*).

I'm glad that I lived in Turkey for a year and began to hear music in a new way. Then when I eventually went to live in China, its traditional music was quite different again. For me, it involved yet another aural overhaul.

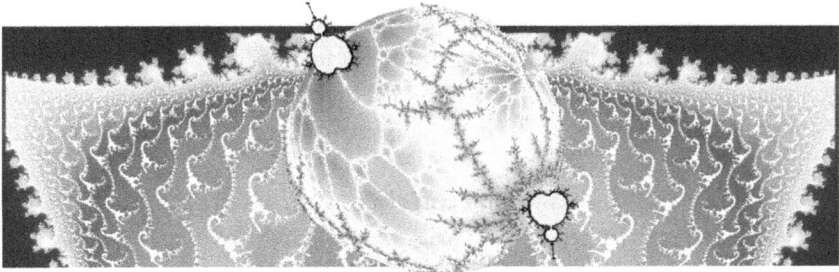

Chapter 11. Dreams Take Me…Somewhere

1. I dream of the bigram ballet

In early September of 1985, I moved with my husband John to Switzerland, where I began attending the Jung Institute in the Kusnacht suburb of Zurich to study Jungian psychology, especially the interpretation of dreams. Soon I was attending classes at the Jung Institute by day; at night I read genetics, physics, and math in the downtown library of the Eidgenossische Technische Hochschule (ETH for short, the Swiss equivalent of MIT).

I kept having a bizarre recurring dream that had started back in Austin in April. Polarized bricks danced in duets in a long, lovely, wordless ballet whose essential dynamic was incredibly old. By October, that symbolism had turned from - and + polarities to yin — — and yang —— polarities. The dance began at downstage left from a performer's view. I was somehow onstage performing in the yin/yin *pas de deux*, yet I was also watching it in the audience.

Then three other couples danced, one pair sequencing after another. All four movements flowed clockwise around all four quarters of the stage. Each quarter showcased its own *pas de deux* that partnered yin and yang by difference or likeness. As the pairs of polarized lines moved through all four sequences, yin found how to hold its own true identity in relationship with yang.

In that exploratory dance, I identified with yin, discovering, "I am this!… and sort of like this…not at all like that…nor exactly like this"…until at last, I returned to the original first position at downstage left, recognizing, "Yes, I am like this! And I know how to dance with yang!"

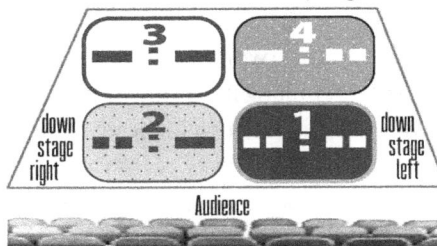

The four bigrams dance as duets of polarity

At that time, I'd never heard of natal hexagrams, nor learned that mine are so very yin. My life's dynamic pattern for the first 36 years was Hexagram 2 ☷☷ morphing into Hexagram 24 ☷☳. I felt that, too, continually. Then came Hexagram 16 ☳☷ morphing back to Hexagram 2 ☷☷, which I also recognize.

2. Looking at the yin background

The 4 bigrams danced the 4 ballet duets. After repeatedly having that dream for 6 months, I wondered why it kept whacking me over the head with a symbolic 2 by 4? Was it telling me to look beyond attention-grabbing yang to yin? To seek an unnoticed proof waiting in yin's humble matter, not in yang's lofty logic? In other words, was the dream prodding me to find an already evident yet mute proof linking DNA's biochemistry with the I Ching's abstract hexagrams?

Then I read Martin Schönberger's work, which mentioned that the I Ching's first title was just *I*, the Chinese word for *Changes*. He thought its brushed glyph looked like a head atop a partial spiral of DNA, signifying intelligence. So Schönberger drew four more rungs on that glyph to produce DNA's full 360° turn. He thought the glyph might be an ancient effort to reference DNA itself.

1. Glyph   **2**. Head + 4 rungs - laid sideways   **3**. DNA's full 360° turn = 8 rungs

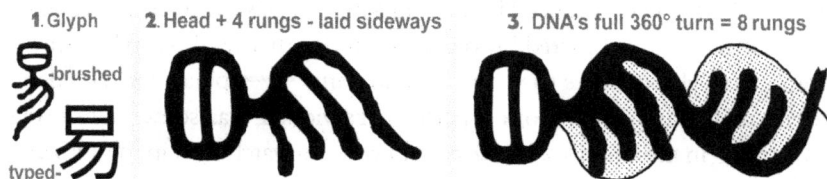

¹*Changes glyph* ²*Glyph of 4 rungs set sideways* ³*DNA 360° spiral = 8 rungs*

Perhaps that old Chinese glyph for the I Ching's early title, *I*, did reference DNA. It was intriguing speculation without corroboration. I had no way to confirm it with any archeological or cultural proof.

But maybe a real archeological object did sit out there somewhere? Something truly physical, extant in mute matter that would speak for itself? Perhaps some little relic still existed that referenced DNA? A trinket…a wall hanging…an old picture? Some forgotten artifact in the historical past?

I searched for a relic originating far back, to when the I Ching first appeared in ancient China. Meanwhile, I also wondered how to look for an object sitting in the yin background without pulling it into the yang foreground of attention?

Yin evidence would be an object that sat in the relational background, not at the logical foreground of notice. Yin symbolizes earth. Unnoticed, silent, fertile matter. Quietly, patiently, its opaque mystery holds, contains, and supports relationships unnoticed. Earth is matter, *prima materia, mater*, mother.

I wondered where to look? Maybe a material object sat mutely dormant, waiting for me to find it...if I looked. Something hidden in plain sight. Finding it could ground the gridwork of abstract theory I'd been constructing with all those coding overlays of hexagrams and amino acid families that I kept drawing with colored pens onto clear plastic sheets, seeking the right fit.

And maybe I was already doing that dance in the dream, already going through its steps to find a historical artifact that indicated the ancient Chinese had somehow found, seen, even acknowledged the I Ching/genetic code correlation...although I certainly could not explain how they might have come upon it, whether by meditative intuition, by an ancient science, or by receiving it from aliens. Pick your method.

Okay, let's get weird—what if the I Ching was proffered from a space ship by an alien intent on showing humans the most important secret of our universe: that the universe itself lives, that we live in its larger, flowing life, that it evolves by a code, that we do, too, and that we can do it more consciously?

3. The He Tu plan & Luo Shu writing

What material artifact still existed from that ancient time? Not much. It took me another two weeks to recognize two relics that had long been hidden in plain sight. China's murkiest legends have handed down two "maps" of dots from long ago. The *He Tu* map of dots originated perhaps around 3000 BCE. The *Luo Shu* map of dots came about 1,000 years later. Those two "maps" were said to explain the origin of the I Ching.

The name of the *He Tu* map means *Yellow [River] Plan*. What plan, I wondered? The name of the *Luo Shu* map means *Lo [River] Writing*.

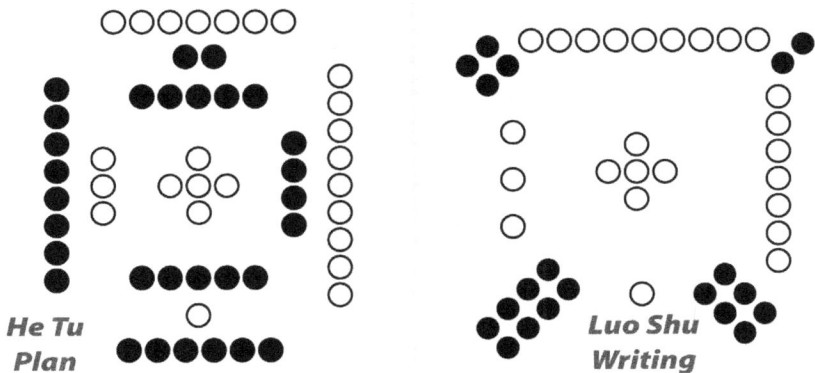

Two Maps: the He Tu plan (left) & the Luo Shu writing (right)

To me, they just looked like black and white dots clustered in orderly groups. Sometimes they're drawn with lines connecting the dots into groups, sometimes not. According to legend, around 3000 BCE Emperor Fuxi saw a dragon-horse rising out of the Ho River. The clusters of spots on its back inspired him to draw a map of black and white dots, said to hold the I Ching plan.

Then nearly 1,000 years later (around 2200 BCE), Emperor Yu, the great flood-control emperor, was inspecting anti-flood construction when he spotted unusual markings on the back of a turtle rising out of the Lo River. The design of spots on its shell led Yu to diagram a new map for the I Ching. Its schematic of dots was said to write out, enact, or command the He Tu's stable plan into active manifestation.

The I Ching, China's oldest classic book, is said to be based on those two "maps" or diagrams of black and white dots. Over many years now, I've made a lot of speculations about those "dots carried on a dragon-horse." They range on a spectrum from the wholly physical to the woolly psychological.

Speculating on possibilities, I've ranged to the extreme of the wholly physical, a completely literal interpretation. Was the I Ching's co-chaos math a gift from an alien culture far advanced beyond Earth's? One may argue that Fuxi would view a spaceship as a dragon-horse. Why not, if it emitted a plume of exhaust or flame like a dragon, and like a horse, it carried things?

That notion, however, only pushed the I Ching's discovery back to an even greater remove in space and time. It did not explain how anyone—human or alien—found the co-chaos math that can shorthand the genetic code, hinting at a master code paradigm rooted in the existence of the universe itself.

Or going woolly, Jungian symbolism would suggest the dragon-horse is a mythic hybrid beast...half divine, half mundane. It combines the creative loft of a yang dragon with the go-power of a horse. A dragon-horse rising from water suggests a creative impetus emerging from the watery depths of the unconscious. A river's two banks control its flow, so that channeled flow is a metaphor for the creativity that Fuxi channeled into making the He Tu map.

Luo Shu Writing   *Magic Square of 15*

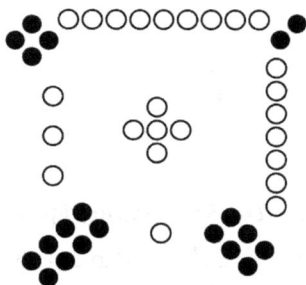

4	9	2
3	5	7
8	1	6

The Luo Shu writing viewed as a magic square of 15

I learned those two maps were honored by rituals that the Chinese had maintained for millennia in their stupendous depth of culture, sometimes forgetting even exactly why. Both maps appear in the I Ching's best-known English translations, those of James Legge and Richard Wilhelm/Cary Baynes.

More recent translations usually omit the He Tu as unimportant. Sometimes the Luo Shu appears, but it is usually presented as just a magic square of numbers, even though the ancients said the Luo Shu depicted the dynamics of "writing out" the He Tu plan...in other words, manifesting that plan into reality.

4. God carries me...somewhere

Nowadays, scholars may look at the Luo Shu's clusters of black and white dots and chide, "Oh, that's just the magic square of 15. If you count each group of dots, you'll see that each row, column, or diagonal adds up to 15."

Yes, that is true. However, it seemed to me that the He Tu and Luo Shu might not be just mere ancient oddities. They might be more significant. The Chinese names for the two maps explicitly state that the He Tu holds the plan, which the Luo Shu writes out. What plan? What writing?

My big question was, "The plan for what? The writing of what?" Sitting at my desk by the bedroom window of our apartment in Zurich's Kusnacht suburb, I studied the two old maps and said, "If the He Tu holds the plan, and the Luo Shu does the writing, how would that parallel the genetic code?"

Oh, of course! DNA and RNA. Ancient China said the He Tu holds the plan; okay, so does DNA. Its double helix carries the genetic plan down across many generations of life. Geneticists even speak of DNA as "the plan."

Ancient China also said the He Tu plan is written into manifestation by the Luo Shu. Modern genetic terminology literally states that RNA writes out DNA's plan by saying that RNA *transcribes* and *translates* DNA's plan into proteins.

So tentatively, I hypothesized that the He Tu dots show the DNA plan, while the Luo Shu dots somehow describe RNA's schematic for turning DNA's architectural plan into living matter. Where and how to begin testing that idea?

First, I tackled the He Tu. I stared at its black and white dots from various angles, seeking correlations with DNA. I was just playing around, you see, relaxing, getting away from ordinary logic. Pondering a copy of a diagram handed down from the oldest I Ching times in ancient China.

Now I am about to show you something quite remarkable. I suspect that the ancient He Tu depicts the actual *atoms* in the four DNA base molecules of Thymine, Adenine, Cytosine, and Guanine. Impossible? Before that recurring dream of dancing led me to investigate it, I would have said so myself.

Judge for yourself. I made this discovery in April of 1986, and its shock propelled me onward through 35 years of reading documents and making worksheets, computer text, and charts into this series you are now exploring.

"What exactly is DNA's basic plan?" I asked myself, sitting at my desk by the window. "Oh! The answer, of course, is the 4 base molecules of T, A, C, and G. They are a polarized pair of pairs. Those 4 molecules encode the genetic plan on the double helix. So how might those 4 DNA molecules parallel the He Tu's mute dots?"

I stared at the black and white dots. Ancient records say the He Tu plan had already been preserved for over a millennium before an accident occurred. According to James Legge, "...the [original] thing, whatever it was, is mentioned in the *Shu* as still preserved at court, among other curiosities in 1079 B.C." But about 100 years later, the He Tu became lost or damaged, so a new copy was recreated from memory by scribes. It passed on down through 3,000 more years to the present day. Chinese archeologists may even someday find a site containing the He Tu map as it looked at the very beginning.

From that story, I wondered if our extant map of He Tu dots, the merely 3,000-year-old copy, might have become more stylized during that long-ago recreation from memory? It seemed likely that the tallies of black and white dots would remain the same—counts and colors are relatively easy to remember—but their general locations would be recalled by approximation.

I also wondered if during that long-ago redrawing from memory, the dots got shifted around slightly or realigned to suit a more stylized taste. After all, things often do get more stylized over time as they become iconic.

On a whim, I simply set the DNA molecules and He Tu dots down together on a worksheet. How? First, I photocopied a diagram of DNA's four base molecules from page 132 of James Watson's *Molecular Biology of the Gene*.

Then below it, I drew by hand the He Tu's groups of black and white dots, and for some obscure reason, I chose to use the diagram of He Tu dots that appears in Da Liu's *The Evolution of I Ching Numerology*. It shows each group of dots and also connects them internally by lines. (The Wilhelm/Baynes translation also has an image of the He Tu with line-connected dots.)

Following you see my old worksheet. On its upper left is the photocopy from Watson's genetics book with the atoms of DNA's 4 base molecules bonded as a polarized pair of pairs—T-A and C-G. Just below it, I drew the He Tu dots, so my eyes could easily scan back and forth between the two diagrams.

The right side was originally empty as my gaze jumped back and forth between the two maps repeatedly...molecules to dots...dots to molecules... and nothing at all occurred to me. The right side of the sheet stayed blank.

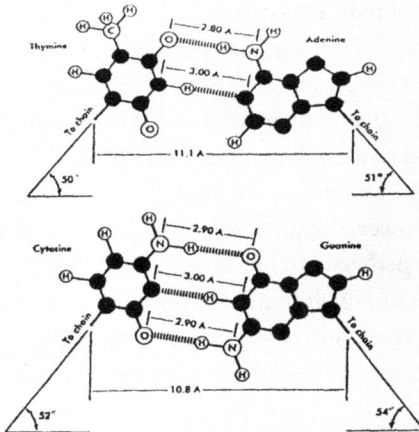

FIGURE 4-14 The hydrogen-bonded base pairs in DNA. Adenine is always attached to thymine by two hydrogen bonds, whereas guanine always bonds to cytosine by three hydrogen bonds. The obligatory pairing of the smaller pyrimidine with the larger purine allows the two sugar-phosphate backbones to have identical helical configurations. All the hydrogen bonds in both base pairs are strong, since each hydrogen atom points directly at its acceptor atom (nitrogen or oxygen).

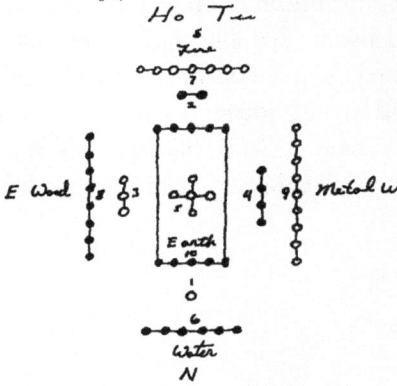

Worksheet of April 1986

But that night, I went to bed and dreamed that God picked me up and carried me...me, an ordinary person!...somewhere. No airplane carried me, no angel, no magic carpet or flying dragon or winged horse...it was God as grand organizing design! Exactly where I went, though, I cannot say. I only knew the trip was wonderful beyond words to tell, and it seemed to go on all night.

I awoke drowsily the next morning in a blissful composure, wondering where God had carried me? I did not know, but it was good. Then I woke up some more and wondered why in the world I would have such a dream?

All morning I looked at my worksheet occasionally as I did tasks in the apartment. But I still had no ideas. The right side of my worksheet stayed blank.

5. Working with split attention

A bit before noon, I folded the worksheet into my purse and walked 14 blocks over to the Jung Institute to hear the first of a week of lectures from a visiting scholar. His theme was the *Divine Coniunctio*…body and soul, conscious and unconscious, extended to a universal level and beyond.

Normally, I paid close attention to lectures at the Jung Institute because they were so interesting. That day, however, I did something odd…at least, odd for me. Sitting in an audience of perhaps 80, I took the He Tu worksheet out of my purse and gave the lecturer only half of my attention.

The other half went down to the black and white dots resting on my lap. One ear, one eye, and one-half of my brain heeded the talk, while the other half studied my worksheet.

The professor lectured—very interesting, too—yet his voice also became a hypnotic drone that burdened my attention and split it into two different focuses…which presented such a heavy sorting task to my brain that the veil beyond my conscious mind thinned enough to let a yang dragon of creative force surge forth from that deeper, oceanic unconscious and fly up into view.

Now, from this later perspective, I assume that splitting my attention into two different information streams acted like a double induction in hypnosis; it let unconscious material rise up and become integrated into awareness.

"Get simple," I said to myself. "View this He Tu plan with the natural, simple mind." Like a child, I just stooped to counting dots on the worksheet.

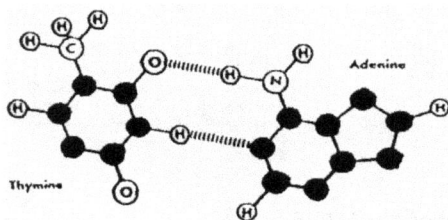

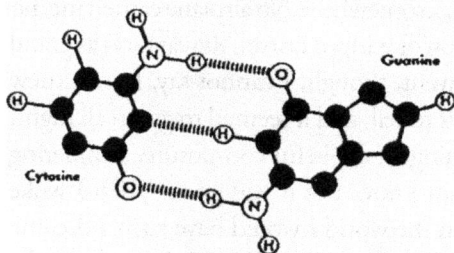

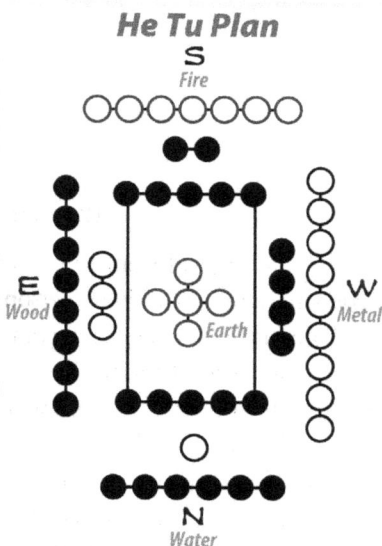

Standard diagrams of DNA base molecules & He Tu dots

As the lecturer's words on the *Divine Coniunctio* buzzed over my head, I looked at DNA's atomic layout and the He Tu plan side by side and intently counted the number of atoms in the DNA molecules: 55! Then I tallied the dots in the He Tu plan: 55! What? I double-checked.

And yet again, it was 55! A shiver went through my body like a wave. How many things in this world can you name that share that very distinctive number of 55? It is an out-of-the-ordinary number, not typical in folklore, which dotes on the transformative 3 and sacred 7, on unlucky 13 and arduous 40.

You can count atoms and dots for yourself. I even did it several more times myself, just to believe what I was seeing. But it was true. What a shock!

Below is my note from the worksheet. I'll decipher my chicken-scratch handwriting: *Note there are 55 atoms in the hydrogen-bonded base pairs. There are 55 dots in the Ho Tu.* (I used the Wade-Giles transcription as *Ho Tu*).

Note 1: 55 atoms in the DNA molecules; 55 dots in the He Tu plan

But you cannot find a note for the thrill that ran down my spine. This curious fit between new science and old map—55 atoms and 55 dots? Previously I'd compared only the *molecular* level of the genetic code structure with analinear math's co-chaos paradigm grown on a dp-tree.

If both atoms and dots had the same count of 55, then did the two systems actually correlate at the *atomic* level? That would mean correlation at a whole new level, different in kind from what I'd already found at the molecular level.

No longer was I viewing my correlation premise as an abstract, logical paradigm. Instead, I was looking at a historically recorded artifact, the He Tu map, whose 55 dots seemed to echo the 55 atoms found in DNA. If that first correlation held true, there would likely be more parallels between both systems.

Stray thought: why did I bother to draw that fancy He Tu version with the connected dots? It would have been easier, quicker, to leave the lines off the worksheet. But those lines bonding dots into groups now reminded me of the lines that bonded atoms into molecules on Watson's diagram.

It gave me a wild notion. Maybe the atomic-bond lines drawn in DNA molecules somehow parallel the He Tu dots bonded by lines? Absurd! But what if the He Tu map did not even originate from a human hand? What if I'd never seen how our modern science draws atomic bonds in molecules?

I considered the various counts of He Tu groups, trying to parallel them with the various counts of atoms—hydrogen, oxygen, nitrogen, carbon. No go!

Okay, what else might work? Color-coding was ancient China's way of symbolizing polarity. Would the DNA atoms somehow polarize in the same ratios as the He Tu's dots of black or white polarity?

I reread Legge's *Introduction*: "The difference in the colour of the circles occasioned the distinction of them and of what they signify into Yin and Yang, the dark and the bright...." Okay, yin is dark earth; yang is bright sky.

I wondered, "Do the 55 DNA atoms polarize like the 55 black and white dots?" My response was to get simple. Look at it with the natural mind. Yin earth holds receptively; yang sky stretches high. In a way, that parallels female and male genitalia. Yin holds receptively; yang reaches out assertively.

I counted the He Tu's black yang dots: 30. Its white yin dots: 25. Total: 55. Okay, if the He Tu's black and white dots mimic DNA's atomic polarity, could I spot now 30 yin atoms and 25 yang atoms in the DNA molecules?

Meanwhile, the lecturer's words on divine union whizzed through my hair and over my ears as if moving beyond the speed of sound. History holds many stories of illumination arising from the unconscious. Creativity comes from inspiration that is breathed into us beyond conscious knowing. You breathe in spirit, and you get *in-spired* by insights stretching beyond the mere data.

I stared at the DNA molecules diagrammed on the sheet, and suddenly I saw it! Black yin holds receptively; white yang reaches out assertively. If I apply that dictum to the atoms in DNA molecules, I observe that some atoms bond receptively into circular rings, while other atoms stick out from it assertively.

Okay, each DNA molecule holds some atoms that group into a central hub or ring. Yin-like, they hold the hub together. It's the same atomic ring that Kekulé envisioned as a snake holding its tail in a circle...which allowed him to imagine the benzene molecule as a circle of carbon atoms, establishing the ring structure for all aromatic compounds in organic chemistry.

And although each molecular hub is a ring of yin atoms, other atoms stick out from the ring in an assertive, yang-like fashion.

I took a felt-tip pen from my purse. Under the river of words pouring from the lecturer about divine union, I inked in every circular hub of receptive yin atoms and counted them up—the photocopy shows my result as black ink atop the hub atoms. First, I counted all the DNA ring atoms: 30. Next, I counted all the black He Tu dots: also 30! Shock! I could hardly believe it!

I counted again. The 30 ring atoms in the DNA molecules paralleled the 30 black yin dots in the He Tu!

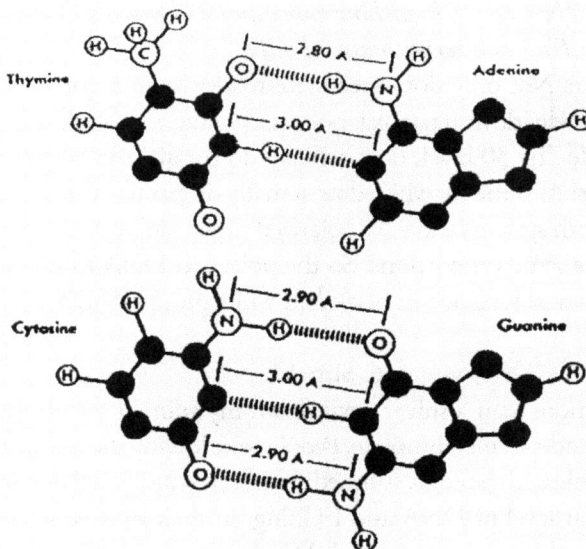

Note 2: black atoms encircle; white atoms reach out

Below is my notation from the worksheet.

Note 3: 30 ring atoms; 30 black He Tu dots

Transcript: *There are 30 aromatic ring atoms and 30 earthly* ● *'s* [black yin dots] *in the Ho T'u - or interior, central units.*

I realized, of course, that in each system, subtracting 30 from 55 would leave 25 for each system...25 atoms branching out like twigs and 25 white dots of assertive yang. But just to be safe and sure, I actually counted each branching white atom, each white yang dot...25 and 25! I did it three times!

Note 4: 25 branching atoms; 25 white He Tu dots

Transcript: *There are 25 branching atoms and 25 heavenly O's* [white yang dots] *in the Ho T'u - or exterior, outward units.*

Holy smoke! Not only does each system add up to a quirky 55 count, but they also subdivide into parallel, polarized counts: 30 black and 25 white each. So the He Tu's 30 black dots actually do parallel the 30 atoms in the molecular hubs! And the 25 white dots actually do parallel the 25 branching atoms! Thus both systems show 55 polarized units, yin or yang, and they do it in exactly the same proportions. So the polarized DNA molecules reflect those dots drawn so long ago in black and white groups! Shock upon shock!

6. Shock!

This time more than a shiver went down my spine. I sat paralyzed! The intensity of it rendered me immobile. People were leaving the lecture hall now, but I sat motionless in the eddy, stunned. Sure, I'd found previous correlations at the molecular level of DNA and I Ching, in their dp-tree structure, co-chaos math, bond-bigrams, unpacked RNA hexagrams, and so on. But now, just by doodling in the lecture hall instead of listening politely, I'd hit upon a new, deeper correlation sitting in mute matter, silent in that old He Tu plan.

Unlike my earlier theorizing about DNA molecules as polarized pairs of pairs, 3-packs, 6-packs, and so on, this dropped down below the molecular level into the atomic level. The He Tu actually mapped the number of atoms in the four DNA base molecules and sorted them into ring atoms and branch atoms.

This was a new kind of congruence between the genetic code and China's most ancient, founding classical document. Both diagrams—atoms and dots—had the same total: 55. Both had the same polarized counts of receptive yin and assertive yang: 30 + 25 = 55. I'd reached into the *ur*-ground of history to find that the He Tu's 55 dots appear to depict DNA's 55 base atoms!

This deeper parallel between DNA atoms and He Tu dots hints that the ancient Chinese somehow came upon the hidden code that generates our bodies and minds, and the I Ching actually records that occasion. Further, it whispers that both systems are two fractal variant codes templated from a deeper master code holding together this card castle made of space, time, matter, and energy.

Wow, that God-dream last night really did carry me somewhere! It dropped me into the mystery of Jung's acausal connection: synchronicity. What are the odds of all this being mere coincidence? Those recurring ballet dreams dancing me through all four quarters of approach? Coaxing me to seek unnoticed yin artifacts in mute matter? Then last night's dream of God carrying me somewhere? And today's lecture of divine union between matter and spirit?

Sitting alone in the lecture hall, my next wonderment was this: "How did the ancient Chinese come upon all this? How did they develop the yin-yang shorthand? How did they tap into DNA's dp-tree structure and codify its math and philosophy, treasure it enough to hand it on down through three millennia? How could such insight way back there jibe with our modern science up here?"

Traditionally, Eastern and Western cultures existed on opposite sides of the globe, and they honored knowledge from very different approaches, right-brain and left-brain. But a data bank hums everywhere beneath our attention in the collective unconscious. It awaits us at the archetypal entrance beneath the ego's focus. It can be tapped in dreams, visions, meditation, prayer. Autistics tap into it willy-nilly. Shamans do it purposefully. And all draw from the same source.

In *Memories, Dreams, Reflections*, Carl Jung said, "Life has always seemed to me like a plant that lives on its rhizome. Its true life is invisible, hidden in the rhizome. The part that appears above ground lasts only a single summer. Then it withers away—an ephemeral apparition. When we think of the unending growth and decay of life and civilizations, we cannot escape the impression of absolute nullity. Yet I have never lost a sense of something that lives and endures underneath the eternal flux. What we see is the blossom, which passes. The rhizome remains."

Now I think of my great-grandmother tucking a bundle of iris rhizomes into that covered wagon before it rolled slowly west to the Texas Panhandle. Some of those white irises are now growing in my backyard. Connections through time.

I got up from the lecture hall chair and slowly walked home. Once there, I tried to describe to my husband what went on during that lecture. The words buzzing past my ears. Dots and molecules dancing in parallel. Polarized counts. He turned over the fish fillets sauteing in the pan, and said, "So what's next?"

"I don't know. I feel dazed." I helped him lay out the late, light lunch by distributing plates, flatware, napkins, drinks. Then we ate, but I sat preoccupied, still juggling atoms and dots in my mind, shifting groups around mentally. Something more still awaited in this correlation. I could sense it. But what?

After lunch, I went to my desk and rotated the clusters of atoms and groups of dots around in different positions, overlaying them in various ways. But nothing worked. Splayed-out atoms did not correlate with formal rows of dots.

All afternoon, try as I might, I could not ride the dragon-horse of creativity any further. I gazed down at my worksheet. Maybe I was carrying this tactic of going stupid too far? I sensed that something still awaited here. Was I taking the right attitude to reach it? On the right side of my worksheet, I queried the I Ching, "Is this going properly?"

Is this going properly?

☷ 51↓   ☳ 54

Yes, I did.

Note 5: "Is this going properly?"

The answer was Hexagram 51 ☳ *Shock!* with a changing Line 2. The name *Shock!* in itself explains this hexagram's electrifying dynamic. Its two trigrams give this hexagram doubled *Thunder.* The Wilhelm/Baynes judgment:

SHOCK brings success.

Shock comes—oh, oh!

Laughing words—ha, ha!

The shock terrifies for a hundred miles,

And he does not let fall the sacrificial spoon and chalice.

Yes, it's true. When I gasped in the lecture-hall after counting the dots and atoms, a real electric shock went up my spine, and a giddy thrill of unexpected success rushed through my body. I can still sense that thrill. Below Hexagram 51, I wrote, "Yes, I did"…affirming I'd felt the shock.

The Wilhelm/Baynes translation also held another mention that pinged me: "The shock that comes from the manifestation of God within the depths of the earth makes man afraid, but this fear of God is good, for joy and merriment can follow upon it." In last night's dream, God carried me on a journey, and then came today's silent wind of words rushing past my ears in the lecture hall.

I admit it…finding this new correlation between atoms and dots did terrify me somehow…because what should I do with it? Gather up my endless notes and actually start turning them into a book? Whether I felt adequate to the task, whether I proved capable of doing it or not, because it was choosing me?

I read further. "Let the thunder roll and spread terror a hundred miles around: he remains so composed and reverent in spirit that the sacrificial rite is not interrupted. This is the spirit that must animate leaders and rulers of men—a profound inner seriousness from which all outer terrors glance off harmlessly."

Finding the parallels between DNA atoms and He Tu dots felt numinous to me, something beyond mere good luck. Maybe those relentless, recurring ballet dreams had prodded me to hop aboard the Tao Midnight Special for the last 6 months, then transfer to a flying dream that lofted me on a nightsea journey toward an illuminating vista of inspiration that I did not anticipate, yet once experienced, its thrill would inspire me to keep on going.

Where was I heading? What was I getting into? I did not know.

Hexagram 51 suggested I should view my work as an offering and "not let fall the sacrificial spoon and chalice." What sacrifice should be continued uninterrupted? Working on this project? To what end? I did feel serious about it, so serious that I now seemed immobilized. I truly didn't know what to do next.

But later in the evening, I picked up the oracle answer, read it again, now thankfully realizing that Hexagram 51 *Shock!* not only describes the shock; its single changing Line 2 also said to stay laid back, don't worry, and wait.

LINE 2: "The second line, divided, shows its subject, when the movement approaches, in a position of peril. He judges it better to let go the articles (in his possession), and to ascend a very lofty height. There is no occasion for him to pursue after (the things he has let go); in seven days he will find them."

Hmm, a position of peril. It nudged a memory from just two days before. I'd met a biologist whose wife studied at the Jung Institute. She suggested that I tell him about the correlations I was investigating between DNA and I Ching math. So I did. He listened for two minutes, looking insulted. He drew up and snapped, "That would reframe biology! Why attempt such a ridiculous thing?"

As I began to explain how the genetic code and I Ching math show evident parallels, he responded by deriding me so thoroughly that his castigating tirade went on for several minutes. I just listened. In a nutshell, he said he doubted there was any correlation, but if so, correlation did not mean causation.

I said, "But I *don't* think one causes the other. I think there's a deeper cause of both, a master code templating many lesser variants...those two among them."

Then he snarled that a university teacher should not overstep. Overstep what? Finally I realized that he seemed threatened by the idea I was suggesting. And his severe, knee-jerk dressing-down had come just two days ago! That response made me wonder...would such hostile attacks be typical?

It forced me to ask, "What am I getting myself into?" If I dared to try, dared to write and publish my co-chaos idea, would scientific minds seek to crush it and me like a moth without even testing out whether it worked?

Was there danger? Real danger? Yes, this study lay far afield from what most Jung Institute students aimed for: becoming a Jungian analyst. Me, I was just following my nose to balance out my psyche's fourth function. I reread Line 2:

LINE 2: Shock comes bringing danger.

A hundred thousand times you lose your treasures

And must climb the nine hills.

Do not go in pursuit of them.

After seven days you will get them back.

Really? So much danger that I will lose my treasures 100,000 times before getting them back in 7 days? It sounded like dramatic hyperbole used to describe needless worry. In analogies, numbers tend to be more qualitative than quantitative. For instance, take "seven days." That timespan is one quarter of the moon's monthly cycle.

It led me to consider how people learned to mark time by watching the cycles of sun and moon. "Many moons ago...." A lunar month has 4 moon phases: new, waxing, full, and waning. It statistically affects monthly menses, epileptic seizures, heart attacks, and insomnia. It causes ocean tides to rise and drop worldwide. Due to eons of moon cycles during our species' evolution, the moon has generated some archetypal memes: menstrual flux, emotional tides, even the Chinese idea that a yin attitude is receptive, reflective, responsive. Since the moon affects our bodies and minds, it influences our cultural heritage.

The moon's elusive network of vibratory associations can make moonlight seem special, romantic, mysterious. Yet moonlight is merely reflected sunlight. Indeed, the moon's yin way of reflecting the sun's yang, projective light is what allows us to see that rocky little satellite orbiting Earth.

Changing Line 2 suggested that I should rise up above the DNA atoms and He Tu dots and reflect on it all. Just let go of its 100,000 details and gain some perspective over time, maybe seven days. Since the number 7 has a spiritual resonance, I would ask for enough strength to keep on going...because I already suspected that I'd find something more, yet not as much as I wanted.

Why did I suspect that? I already knew the I Ching well enough to have noticed that changing Line 2 would morph Hexagram 51 ☳☳ *Shock!* into Hexagram 54 ☳☱ . Wilhelm/Baynes calls this hexagram *Marrying Maiden.* I call it *Delayed, Diminished Union.* It's a marriage, but not an equal one. Its dynamic did not bode full closure ahead on this He Tu mystery, only a partial solution. So I decided not to read Hexagram 54 tonight. It might prove too dispiriting. Right now, I'd just try to let go of my tension and get above it all.

"Relax," I said. "Let the shock wear off. Take a break. Talk. Eat. Sleep."

And I found enough strength to go on. Over the next 5 years, I slowly began to grasp how the 4 bigrams grow on a p-tree, how they code for DNA's base molecules (T, A, C, and G), how *bond*-bigrams code for the 64 DNA 6-packs in 64 hexagrams, how *ordinary* bigrams unpack RNA codons into the 64 RNA message hexagrams, how they automatically sort into their amino acid families, and how each amino acid's task even syncs up with its hexagram's philosophical dynamics.

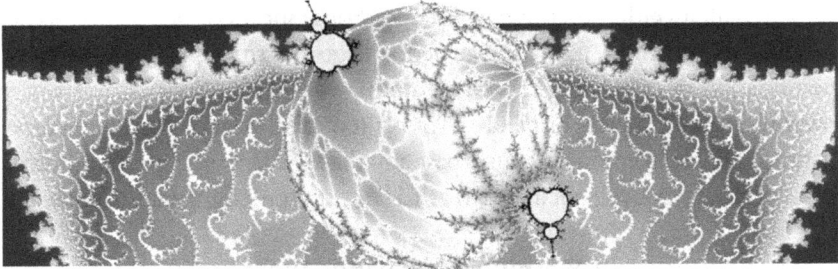

Chapter 12. The Atomic Map

1. I anticipate a delayed, diminished union

That night I went to sleep and again dreamed that God was carrying me… somewhere. But exactly how or where seemed even vaguer than on the previous night. When I woke in the morning, that nightsea journey felt wonderful again, but without the simple clarity and joy of its previous night's passage.

As I ate breakfast, I decided this new dream felt uplifting, yet its unresolved quality left something in mystery. Uh-oh! Was this dream reinforcing what I already suspected yesterday upon realizing that Hexagram 51 morphed into Hexagram 54? Was the dream, too, warning that I would not fully resolve this atoms-to-dots mapping? Oh well, I'd keep working on it anyway, taking it as far I could muster, even if some correlation aspects remained shrouded in mystery.

That morning I sat at my desk and pondered Hexagram 54, hoping its dynamic would speak to me more clearly, more fully, and especially, more hopefully. I started researching the hexagram's background. Its central analogy describes an arranged marriage for a young maiden of high birth.

Hexagram 54 recounts the ceremonial procedures of bringing a bride of royal birth into a groom's rural, impressed but less distinguished family. Historically, this royal bride was probably Tai Jen. Around 1055 BCE, she wedded the father of Ji Chang. Her son was the man who'd eventually be imprisoned, write down the first I Ching document, and later become known as King Wen.

Recall, this anecdote came from a feudal, polygamous culture from more than 3,000 years ago. Its main point is that even after all the protracted rituals, this high-born girl still won't be the first wife, but instead, only a younger wife who must act with caution and reserve. Thus this union is not only delayed, but it is also diminished, i.e., it is less than she would wish for. I call this hexagram *Delayed, Diminished Union* because to me, it describes that fractal's dynamic well.

Last night I threw up my hands in frustration at this response to my query on advancing the atom/dot correlations. But not today. Today I just accept the quest. It may involve much effort, procedure, delay, then yield a result that, if substantial, is less than fully satisfying, But okay, I'll continue.

After all, it's either that or give up now. Some success is better than none.

So this morning, I'm forewarned. I'll go on through every possible permutation of correlating the DNA atoms with He Tu dots. I'll not worry overmuch about success or failure. Not get too attached to a 100% outcome. Instead, I'll just keep on chugging past any difficulties. Some good will come of it eventually. Maybe not as much as I'd like, but something fairly significant.

2. Warning: birth pangs ahead!

Imagery in the I Ching is meant to be studied both logically and also associatively. It calls up webby networks of thought that weigh the resonant options in a hexagram's dynamic. The trick is to choose the best among those options, moving on through its dynamic as well as possible. I knew that much, but I still felt lost on exactly where or how to begin.

I decided to prime the pump of inspiration with yet another query of the I Ching. Beneath Hexagram 54, I wrote: *"The nature of this 54?"*

Note 6: "The nature of this 54?"

Why ask that question? I wanted to know more about what to expect as I made my way toward a delayed, diminished union indicated by Hexagram 54. Basically, I was saying, "I want more information on how to let this fractal pattern carry me on through any setbacks to the best possible potential."

The answer was Hexagram 3 ☳☵ *Initial Difficulty* or *Birth Pangs*, with only one changing line, Line 1. In Volume 3 of this series, the last chapter examines Hexagram 3 in detail. The dynamics of Hexagrams 1 and 2 act as father and mother to what happens in Hexagram 3. It arises from their relationship, so Hexagram 3 describes the dynamic of giving birth. In it, a difficult beginning results in the successful birthing of...something.

Giving birth, I knew, meant an organic struggle to deliver something new. A plant. A baby. A company. A nation. An idea. To inform and reinforce my resolve, I read various commentaries and opinions on Hexagram 3.

Of Hexagram 3, Legge said, "there will be great progress and success, and the advantage will come from being correct and firm. (But) any movement in advance should not be (lightly) undertaken. There will be advantage in appointing feudal princes." For me, "appointing feudal princes" suggested reading various authorities. Legge said my changing Line 1 "shows the difficulty (its subject has) in advancing. It will be advantageous for him to abide correct and firm; advantageous (also) to be made a feudal ruler." I decided that being made a "feudal ruler" meant to trust my own judgment in testing and evaluating the various options.

In short, it'll be hard for me to advance, so I should stay correct and firm in it, not taking it lightly. "Delay and difficulty ahead," I thought. "So what? I'll persevere." As to being a ruler, I'll follow my nose on this. After all, talking to that biologist two days ago had just annoyed him mightily. I jotted…

Note 7: I vow to give it my best try

Transcript: *3 Difficulty in the beginning has supreme success. Great progress & success from being correct & firm.* About Line 1, I wrote, *Difficulty in advancing. Be correct & firm - - necessity of caution and of taking authority. Okay*

3. Holding Together

Hexagram 3 ䷂ has changing Line 1 that morphs it into Hexagram 8 ䷇ *Holding Together*. This is one of my favorite hexagrams. It says holding things together brings benefit. Its trigrams put yin's abiding *Earth* ☷ below, and swift, channeled *Water* ☵ above. As water flows, it cuts routes into the earth. They connect and stream toward the oceans. On our planet, water holds things together at every scale—from the spherical tension of a dewdrop, to the fluid plumping up your body, to the bright blue ball holding continents in its shining blue net when viewed from space. Here is Wilhelm/Baynes *Judgment*:

HOLDING TOGETHER brings good fortune.
Inquire of the oracle once again
Whether you possess sublimity, constancy, and perseverance;
Then there is no blame.
Those who are uncertain gradually join.
Whoever comes too late
Meets with misfortune.

I did not want to baulk and delay until I joined in the dynamic too late, losing my opportunity, so I took the plunge. For the first time, I jotted down my unspoken hypothesis. Even writing down the actual words felt daring.

Note 8: Hypothesis

Transcript: *Hypothesis: Ho Tu. This is a stylized schema of the H-bonded base pairs of DNA, rearranged for mnemonic ease of retention over time.*

Daring, yes! Remember, this was in March of 1988, back before I'd written any of these books. I was still in the idea-gathering, note-taking, diagram-drawing stage. Moreover, I knew of no scholarly precedent for my proposal that the He Tu is a stylized rendition of the atoms in DNA's four base molecules.

I'd seen no word of this idea in a research paper, nor heard any talk about it…no mentor, class, or colleague who championed the notion. Nor was there anyone here I could discuss it with, except my husband, who just smiled at me, puzzled. So I went far out on a limb…

…that turned out to be just the first of many. Little did I realize I was embarking on more than 35 years of exploration into much wilder territory than this He Tu conjecture, on toward the end of the series…or as much of it as I can finish. I have ideas for 8 volumes, but let's see how far I actually get.

Next, as Hexagram 8's *Judgment* suggested, I queried the I Ching on whether I had enough "sublimity, constancy, and perseverance" for this task. The answer was Hexagram 40 ䷧ *Deliverance* with no changing lines. Its analogy describes slaves who are freed to go back to their former homes. They have a new appreciation for freedom now regained. This dynamic returns to an old condition, but it's now seen with a fresh, appreciative perspective.

I could see how the dynamic of deliverance applied to my situation. Yes, correlating the 55 DNA atoms and 55 He Tu dots did free me from doubt by reinforcing my original hypothesis that the genetic code and I Ching are based on the same paradigm. Now I'd even found evidence at the atomic level in the He Tu map of 55 dots, a historical document known in China for over 3 millennia. I held to my old condition of testing the original premise while now also appreciating its deeper-level record of atoms/dots in an old document.

I confirmed aloud, "I'll hold onto my purpose in testing the correlations, despite difficulties that arise, despite a less than full success." Over the next 4 days, I opened books, examined atomic diagrams, studied the He Tu. I drew many charts. I talked with my husband John. He would listen with his finger stuck into a book to hold his place. It reminded me of the little Dutch boy story, but this time, patiently trying to stem a relentless flow of words, not water.

I attended the week's remaining *Divine Coniunctio* classes at the Jung Institute, but as my eyes traveled back and forth between atoms and dots on the worksheet, nothing new came of it…except wondering if maybe during that historical recreation 3,00 years ago, some ancient scribe's failing memory or his purposeful intention to stylize had shifted any of the He Tu dots into a different group. Maybe he (surely a *he*) even dared to tidy it up a bit while redrawing it, corralling those dots into a more orderly, formal, sequential design.

4. I find a polarized pair of pairs in each 55-part plan

"Be patient," I reminded myself. "The moon's monthly cycle has four quarters, so I'll need to see this from a new quarter of contemplation…."

Oh! The DNA diagram is divided into four quarters…its 55 atoms group into four molecules. They sit in four *squared-off* quadrants on the map!

So does the He Tu plan also show four quadrants? Yes! It puts the dots into four *triangular* quadrants—as South, North, West, and East."

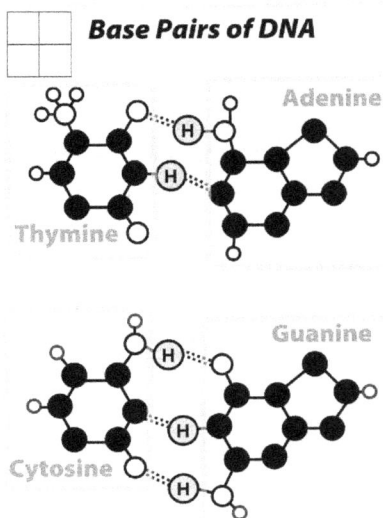

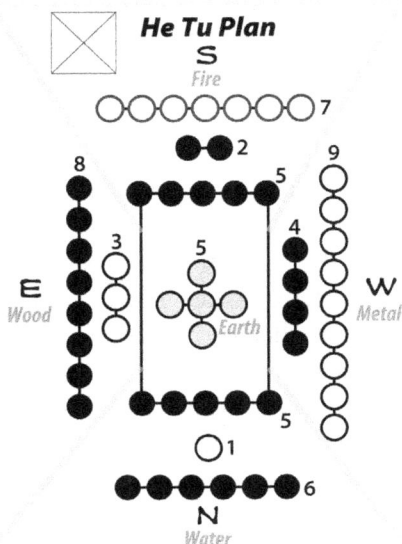

Each plan has four quadrants as a polarized pair of pairs

Note that ancient Chinese maps set South at the top, North at the bottom, East on the left, and West on the right...the exact opposite of Western maps.

I thought, "Okay, I see four quarters in each system. How do they correlate? Oh! Each system is organized into a polarized pair of pairs! The four He Tu quadrants polarize into the N-S axis and E-W axis. The four molecules polarize into the T-A pair and C-G pair. So both systems put 50 of their 55 units into a polarized pair of pairs! Here is yet another parallel in symbol distribution!"

5. I find 5 central connectors in each polarized pair of pairs

Then I realized Watson's diagram puts 5 hydrogen atoms at the center of DNA's base molecules. I inked them in with a green felt-tip pen. (See next page) The 5 H atoms bond together all 50 other atoms in the T-A pair and C-G pair.

Likewise, the He Tu diagram puts 5 dots in the center. I inked them in with the pen. They bond together all 50 other dots on the N-S axis and E-W axis.

Therefore, each system has 5 central connectors that bond together its 50 other units, which are polarized into a pair of pairs!

Here was yet another parallel! These parallels at the atomic level brought those earlier parallels that I'd found at the molecular level into a wider congruence. And this increasing number of parallels at multiple levels also reinforced my premise that both the genetic code and I Ching math are fractal variants on a deeper co-chaos paradigm. I was elated. Already I'd found so many parallels between atoms and dots, and there might be even more!

It is important to note that although I'd darkened the He Tu's 5 central dots, they are white on the ancient He Tu plan. Indeed, for over 3,000 years, Chinese scholars have fretted over a paradox: why are the He Tu's 5 central dots colored white? Normally, naming them *Earth* should mean they are colored black to signify yin. Nevertheless, on the historic He Tu map, the 5 central dots are shown as yang white, yet they're also named yin Earth! This is a 3,000 year-old puzzle!

6. At play in the quadrants

Next, I thought, "If the 5 central white dots actually do symbolize the 5 H atoms, then perhaps each quadrant of dots stands for a specific molecule? If so, which quadrant of He Tu dots stands for which quadrant of atoms?

Hmm, maybe they're paired along the North-South/East-West axis...or maybe they sit in adjacent quarters like directions on a compass...or what else?

I tried 18 options over a full month of effort. I investigated constructive and destructive cycles of the Chinese elements, the sugar-phosphate backbone of DNA, etc. I will not tire you with every permutation and ramification I tried. Rather than describe those protracted procedures, I will summarize it thus:

First, find the total count of DNA atoms in each squared-off quadrant.

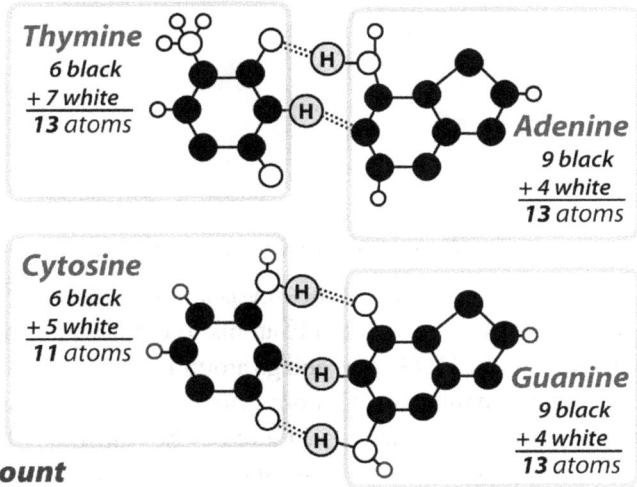

Thymine
6 black
+ 7 white
13 atoms

Adenine
9 black
+ 4 white
13 atoms

Cytosine
6 black
+ 5 white
11 atoms

Guanine
9 black
+ 4 white
13 atoms

DNA atoms count

DNA atoms count

Next, find the total count of He Tu dots in each truncated-triangle quadrant.

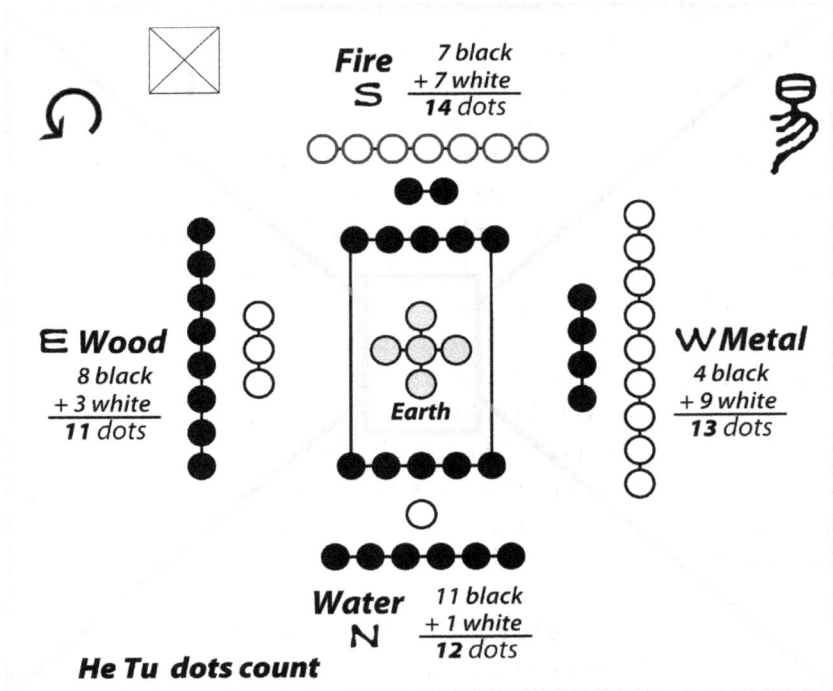

Fire
S
7 black
+ 7 white
14 dots

E Wood
8 black
+ 3 white
11 dots

Earth

W Metal
4 black
+ 9 white
13 dots

Water
N
11 black
+ 1 white
12 dots

He Tu dots count

He Tu dots count

DNA atoms minus ↻ 5 central atoms	He Tu dots ↺ minus 5 central dots
Thymine–13	Wood–11
Adenine–13	Water–12
Guanine–13	Metal–13
Cytosine–11	Fire–14
Total: 50	**Total: 50**

Note 10: All quadrants of each system = 50

Finally, compare their totals, using only the quadrants of 50 atoms/50 dots. Ignore for now those 5 central H atoms/5 central white dots. Start at the top left corner of the DNA chart and go around it *clockwise*. Find each quadrant's atom total. Thymine has 13 atoms. Adenine has 13 atoms. Guanine has 13 atoms. But Cytosine has only 11 atoms. The sequence is 13, 13, 13, 11.

However, on the He Tu chart, start at the top left corner and go around it *counterclockwise*. Wood in the east has 11 dots. Water down in the north has 12 dots. Metal in the west has 13 dots. Fire up in the south has 14 dots. Its counterclockwise sequence counts out a tidy progression of 11, 12, 13, 14.

In comparison, three DNA quadrants have 13 atoms each, leaving a single quadrant with only 11 atoms...or 2 atoms less. By contrast, the He Tu's four quadrants placidly count out a simple progression of dots per quadrant: 11, 12, 13, 14. That numerical slippage morphs it into a smooth counting sequence.

Yes, the total is the same in both systems: 50. Fine. Yes, both systems also split into 30 black and 25 white. Again, fine. But no matter how I flipped around the quadrants, no matter how I torqued the four molecules, studied their sugar backbones, considered the DNA 6-pack versus the RNA three-pack, twiddled Chinese elements, etc....no matter how hard I tried to seek a perfect fit between atoms and dots per quadrant, nothing matched precisely. In every option, the allotment of dots and molecules differed per quadrant.

Yes, each system totals 50 units in its four polarized quarters. Moreover, each also has 30 black and 25 white items. Further, each shows 5 central bonds. But even today, I still cannot find a perfect fit between atoms and dots per quadrant. No matter what, the count stays off on a quadrant-by-quadrant basis...so as Hexagram 54 predicted, my final satisfaction is only partial.

7. Why are the He Tu's 5 central dots called White Earth?

I suddenly realized why the ancient Chinese may have viewed those central 5 dots as both *white yang* and *Earth yin*...because, look, the 5 central white dots really do combine yin and yang features. Those 5 white dots sit in the center to bond together the 50 other dots (holding things together is a yin trait). Yet the 5 white dots also do it by reaching out to each different quarter (stretching out assertively is a yang trait).

Likewise, the 5 H-bond atoms do much the same thing. They hold the 50 other atoms together (a yin trait), yet they also do it by reaching out to each different molecule (a yang trait). Thus, the 5 H atoms act both yin and yang; their function really does parallel the He Tu's 5 central white Earth dots.

In other words, the He Tu's 5 central dots mimic the behavior of the 5 central H atoms. That would explain why they are colored white yang, yet they are also named yin Earth. I felt satisfied that it resolved, at least for me, an age-old mystery for Chinese scholars in a paradox of 3,000 years duration!

Maybe Fuxi saw 5 hydrogen atoms bonding together four molecules—in a dream, meditation, or on a diagram handed to him by aliens slithering out of a spaceship? I don't know. In any case, each system has 5 central units...atoms or dots. They bond together a polarized pair of pairs made of 50 other units... atoms or dots. In each system, its 50 units sub-divide into...30 atoms in ring molecules and 25 branching atoms...or into 30 yin dots and 25 yang dots.

To me, this multi-level correspondence is far beyond random chance. It also hints that both systems are two lesser variants of a deeper master code of polarized information hidden in the spacing and timing of all matter and energy. To me, this approaches the Divine Coniunctio of universal union.

Such congruity in so many ways leads me to wonder if scribes stylized the He Tu dots into orderly rows when they recreated it from memory over 3,000 years ago, "improving" upon its original layout from perhaps 4,000 years ago.

Perhaps the "damage" to the original manuscript was even contrived to justify "fixing" it. So many other parallels do exist between the atoms and dots that this sequential dot count of 11, 12, 13, 14 may actually be a tip-off to an ancient "map doctoring." At least, that is my best explanation for the disparity between the count of atoms per molecule and the count of dots per quadrant.

But maybe that is not the answer. Perhaps a deeper reason exists. It may even reflect a literal slippage of two atoms in switching from Thymine in DNA to Uracil in RNA. I cannot fathom how, though. Or something else. I don't know.

Nevertheless, my month of investigation led me to suspect that the He Tu plan, the DNA atoms, and perhaps even the polarized beats that drum up space and time, prepping it for matter and energy...all use a variant of the I Ching's counting-out algorithm in its ancient yarrow stalk ritual.

Why? Because the ritual begins with 50 stalks...as if all 5 central H atoms are already taken out of the count, leaving just 50 atoms to deal with. Then at 50, another yarrow stalk is immediately set aside and never again touched during the many calculations done with the other 49 stalks. Perhaps that initial loss of 1 stalk/atom triggers a dynamic for the other 49. Again, I don't know.

In another odd echo, both pyrimidines, T and C, have molecular rings of 6 atoms each, but both purines, A and G, have molecular rings of 9 atoms each. Hmm, 6 and 9. They echo the ancient yarrow stalk algorithm where 6 indicates changing yin, while 9 indicates changing yang. Again, congruence whispers here.

To my mind, the multi-level parallels across both systems show correlations in 6 ways: (1) The 64 hexagram 6-packs shorthand DNA's 64 molecular 6-packs. (2) The 64 RNA codons unpack into 64 RNA message hexagrams sorted into their amino acid families. (3) The RNA hexagram messages reflect their amino acid traits. That's all molecular. (4) At the still-deeper atomic level, the 55 DNA atoms mimic the 55 He Tu dots. (5) Schonberger pointed out that the Chinese ideogram for *change* mimics the DNA double helix. (6) Both systems share the co-chaos math paradigm grown on a dp-tree.

These correlations observed from 6 different angles seem to me like sextuple-entry bookkeeping. So many parallels from so many angles of investigation indicate, at least to me, that the genetic code and I Ching are two fractal variants of a still-deeper code hidden in universal nature itself. Thus, I postulate that if I Ching math can shorthand the genetic code, and if both variants describe events in matter and energy over space and time, they are templated from a primal master code that operates in mind and matter.

My originating dream began with a candelabra on a mirrored table, foreshadowing the dp-tree rooted in nature. Does nature start at the quantum level that emits blips of matter held together by waves of energy, i.e., particle-waves? No, according to my dream, it originates at the ultra-tiny mobic scale, where pulses of information spin the finer, smoother fabric of space and time.

I cannot explain how the ancient Chinese silhouetted the 55 atoms of the DNA molecules in the 55 dots of the He Tu plan. Did it emerge from daydreaming, as with Kekulé's snake? From a meditative insight by some savant tapping into the *ur*-ground of math? From a lost civilization such as Atlantis or Mu? From a pit stop made on this planet by aliens journeying to elsewhere, or perhaps shipwrecked here and making their missionary best of it?

Or is it just sheer chance that the genetic code and I Ching reveal so many correlations from so many angles of approach? Can chance get this sheer and slick and hard to dent? What do you think?

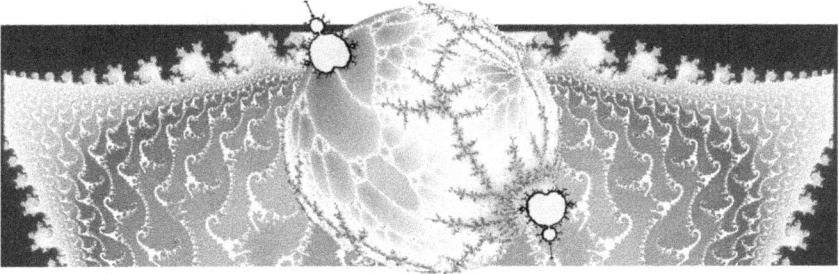

Chapter 13: Lost Worlds

1. Two lost "books" of hexagrams

Long ago, there was not just one I Ching book, but rather, three. They are mentioned in the *Rites of Zhou* (Chou). It says the Grand Diviner was in charge of three I Ching books, and each book held 64 hexagrams. The first was the *Lian Shan* (*Change Manifests in the Mountains*) from the Xia dynasty (2070–1600 BCE). The second book was the *Kuei Tsang* (*Flow & Return–Womb to Tomb*) from the Shang dynasty (1600–1046 BCE). The third book was the *Zhou Yi,* written just before the Zhou dynasty started (1046–256 BCE).

The first two books are now lost except for fragments. Records are scanty on what they contained. According to James Legge in *I Ching: Book of Changes,* "In each book, the regular or primary lineal figures were 8 [trigrams], which were multiplied in each till they amounted to 64 [hexagrams]. A few texts from the Han dynasty (206 BCE–220 CE) describe them briefly, stating that *Lian Shan* had 80,000 words and *Kuei Tsang* had 4,300 words."

Some say those lost books were not actually what we call "books" today. In the past, a book was often a much shorter affair. Perhaps they were more like three essays containing three different orders of the 64 hexagrams. Of course, varying the hexagram order can vary the coding result, as we saw for DNA and RNA in Chapter 5.

Despite its veneration of scholarship, China also had a history of destroying books when a new ruler considered something in the library politically or intellectually dangerous. The Chinese ability to protect manuscripts rates higher than in most ancient cultures (exceeded only by Tibet), but an occasional purge has brought disaster to both libraries and scholars.

For example, a brash 13-year-old ruler in the western territory of Chin (Qin in Pinyin) began in 246 BCE to conquer neighboring states "like a silkworm devouring a mulberry leaf," to quote the words of ancient historian Sima Qian. Over 20 years, that conquering boy ruler subjugated the Warring States territory, turning it into the Chin empire, and he was its first emperor.

That bold, imperious man banned the use of his old name. His new name? Chin Shi Huang Ti (literally, China's first emperor). His goal? To put an end to every kind of dissension and disharmony. To do so, Chin standardized laws, writing, coinage, weights, measures, and wheel sizes. He built a network of roads and canals. He consolidated and extended the Great Wall of China.

Chin also killed, suppressed, or destroyed whomever, whatever he deplored. He executed over ⅓ of China's population. He banned and burned any books that he felt threatened his reign. Luckily, King Wen's I Ching text was considered useful, so it escaped the great book burning. According to James Legge, "In the memorial which the premier Li Se addressed to his sovereign, advising that the old books should be consigned to the flames, an exception was made of those which treated of 'medicine, divination, and husbandry.' The Yi [*I Ching*] was held to be a book of divination, and so was preserved."

Despite many assassination attempts, Chin ruled until 210 BCE, when he died, probably due to taking medicinal doses of mercury to become immortal. Extravagant emperor that he was, Chin had already prepared a huge tomb of about 20 square miles (55 square kilometers) under Mount Li near Xian.

Chin was buried at Mt. Li with life-size terra-cotta statues of over 8,000 soldiers, 670 horses, and 130 wooden chariots. Excavation of that area began in 1974, to the wonderment of archeology, but most of it remains buried. Someday further excavation of that tomb at the north foot of Mt. Li may even reveal bits of those two lost I Ching books, along with other relics.

Although Chin unified China and standardized many of its values, the Chin Dynasty itself faded away within 4 years after his death.

2. Silk Text: trigram & hexagram orders

A new order of hexagrams turned up in the 1970s. The excavation of a 168 BCE tomb in Hunan Province found some documents written on silk, so they were together dubbed the *Silk Text*. One of those documents revealed an unusual hexagram order found nowhere else.

We know that the *Silk Text* I Ching is neither of the two lost books mentioned in ancient records. Its text includes hexagram meanings, but a few diverge from the *King Wen* text. It also offers no divination instructions, raising a conjecture that perhaps this version was not even intended for that purpose. Greg Whincup discusses it extensively in *Rediscovering the I Ching*.

The hexagram sequence found in the Silk Text is unique. Its trigrams use gender identities like Shao Yong's Early Heaven, binary order, yet they're sequenced differently. Four males come first: Father, 3rd son, 2nd son, 1st son. Then come the females: Mother, 3rd daughter, 2nd daughter, 1st daughter.

1	2	3	4	5	6	7	8
White Heaven	Purple Mountain	Blue Abyss	Yellow Thunder	Black Earth	Red Lake	Orange Fire	Green Wood-Wind
☰	☶	☵	☳	☷	☱	☲	☴
Father	3rd Son	2nd Son	1st Son	Mother	3rd Daughter	2nd Daughter	1st Daughter

Silk Text trigrams use the Early Heaven genders, but in a new order

This trigram order is put in play throughout the Silk Text hexagram chart below. Its first column has hollow white numbers. Observe how the 8 trigrams march down the first column in this order, and they're doubled into hexagrams!

Hexagram Order of the Silk Text

1							
1	2	3	4	5	6	7	8
2							
3	1	2	4	5	6	7	8
3							
5	1	2	3	4	6	7	8
4							
7	1	2	3	4	5	6	8
5							
2	1	3	4	5	6	7	8
6							
4	1	2	4	5	6	7	8
7							
6	1	2	3	4	5	7	8
8							
8	1	2	3	4	5	6	7

The Silk Text hexagram order

3. Silk Text– 1st column of hexagrams & all top trigrams

THE NEXT TWO SECTIONS EXPLORE THE SILK TEXT'S UNUSUAL TRIGRAM ROTATION MORE THOROUGHLY. IF THAT SOUNDS TOO TEDIOUS TO PURSUE, YOU MAY WANT TO SKIP ON AHEAD TO SECTION 5.

Hollow white numbers sit before the first column of hexagrams. This column's sequence stacks 4 male trigrams that are doubled into hexagrams: ⒈ *Father Heaven,* and then the sons from youngest to oldest—⒉ *Mountain,* ⒊ *Water,* and ⒋ *Thunder.* Below them sit 4 female trigrams also doubled into hexagrams: ⒌ *Mother Earth,* then the daughters from youngest to oldest—⒍ *Lake,* ⒎ *Fire,* and ⒏ *Wood-Wind.* They sit in Early Heaven gender order.

(Note: these numbers do not appear in the Silk Text or in Greg Whincup's *Rediscovering the I Ching.* I merely added numbers here to clarify this unusual hexagram order's pattern of rotating trigrams.)

The defining first column of doubled trigrams makes this Silk Text order of hexagrams unique. It establishes all 8 *doubled* trigrams in the first column, plus it will also dictate the *top* trigram of every hexagram throughout the chart.

Here's how it happens. In a Silk Text chart of hexagrams, the first column's top trigram on each row will dictate every top trigram along that whole row. Thus the same *top* trigram will iterate along its row. However, a binary-order chart of hexagrams will instead iterate the same *bottom* trigram along a row. This means the Silk Text chart holds echoes of Shao Yong's binary order, yet it also pings some aspects of the RNA message hexagram order.

4. Silk Text–bottom trigram rotation

The Silk Text order propels all bottom trigrams on a rotation path moving across and down the chart's rows. Solid (red) numbers indicate the rotation path on this hexagram chart. Four poetic sentences in the *Ta Chuan* (Great Treatise) describe its unusual method of rotating trigrams. Liu Dajun of Shandong University translates it thus: Heaven and Earth establish their position. Mountain and Lake interpenetrate their chi. Fire and Water overcome each other. Thunder and Wind influence each other."

To decipher those poetic nature descriptions, let's see what the four sentences imply. Poetic description can be found in many ancient Chinese texts, and typically it uses analogies drawn from nature. For instance, *Art of the Bedchamber,* a Tang dynasty equivalent of the *Kama Sutra,* refers to the phallus by such nature imagery as the Jade Root, the Male Peak, the Celestial Stem. It calls the vagina the Precious Flower, the Golden Gully, the Dragon Jade Gate.

The *Ta Chuan's* four sentences of poetic nature description hold directives:

A. *Heaven* and *Earth* establish their position." This sentence decrees that the bottom trigram in the first hexagram on the first row is *Heaven* **1** ☰ . Then beside it on the right sits *Earth* **2** ☷ ; this is the bottom trigram of the second hexagram on that first row.

B. "*Mountain* and *Lake* interpenetrate their chi." This sentence puts the bottom trigrams of *Mountain* **3** ☶ and *Lake* **4** ☱ beside each other on the first row to "interpenetrate their chi"…in modern terms, to blend their energies.

C. "*Fire* and *Water* overcome each other." Here the two energies of *Fire* and *Water* are so strong that they "overcome each other," and thus, those two bottom trigrams switch places into sitting together as *Water* **5** ☵ and *Fire* **6** ☲.

D. "*Thunder* and *Wind* influence each other." These are the final two bottom trigrams on the first row. *Thunder* **7** and *Wind* **8** ☴ settle into that first row to sit side by side, influencing each other.

This establishes the first row of bottom trigrams in number sequence of **1, 2, 3, 4, 5, 6, 7, 8**. In all succeeding rows, only the bottom trigram in the first column is displaced in each row's number sequence of **1, 2, 3, 4, 5, 6, 7, 8**.

The result is an intricate yet harmonious mesh of odd-even numbered dynamics. The rotation gear below indicates the odd-even number distribution.

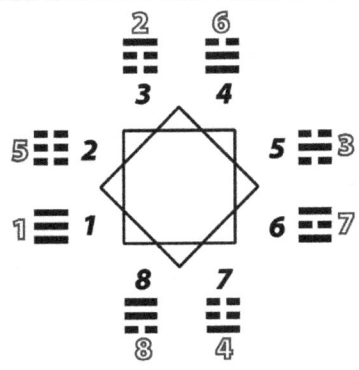

Silk Text decoder

I adapted this decoder diagram from Liu Dajun's *A Preliminary Investigation of the Silk Manuscript Yijing*. Once you spot it, the rotation path of bottom trigrams is simple enough. The first column sorts the solid (red) numbers into odd-numbered male trigrams, then into even-numbered female trigrams. Past the first row, each successive row brings a new bottom trigram to the front of its row, displacing it from that row's standard sequence of **1, 2, 3, 4, 5, 6, 7, 8**.

But what is the point here? Why were Silk Text hexagrams organized in this fashion? I don't know.

5. The King Wen order of hexagram
Consider the King Wen order of hexagrams found in I Ching oracle books.

The King Wen order of hexagrams

Chinese history says the King Wen order was written down by a duke back in the days when he was called Ji Chang, still imprisoned by tyrant King Zhou, the last ruler of the Shang dynasty.

Ji Chang died ten years after being freed. His second son, Wu Wang (Warrior King), established the Zhou dynasty. He gave his dead father, Ji Chang, the posthumous title of Wen Wang (Scholarly King), thus officially making his already-dead father the first ruler of the new Zhou dynasty.

6. Some structural symmetries in the King Wen order

The King Wen order of hexagrams does not sit in binary sequence. Instead, the hexagrams are arranged as 32 pairs sequenced along the rows by their philosophical meaning. Each twosome engages in comparison/contrast dynamics. A philosophical story propels their progression along from procreation to birth, then on through a series of maturing stages of life to death, and then to a paradoxical renewal. Thus the King Wen order holds a philosophical symmetry that pays heed to the major issues in life.

Although philosophical meanings clearly sequence the King Wen chart, this order nevertheless also displays some notable structural symmetries. Mathematically speaking, the comparative polarities in each pair of hexagrams may differ in three distinct ways. In any given pair, their yin-yang lines may *flip-flop* each other's polarity, *balance* each other's polarity, or both *flip-flop and balance* each other's polarity. To break it down in more detail…

On the chart of *King Wen's Hexagram Pairs*…

- **A–24 pairs of hexagrams flip-flop each other's polarity**

In the chart's legend, these are the **A–24 No-Box Pairs**. In these 24 pairs, the first hexagram will upend or stand on its head to create the second hexagram. For instance, Hexagram 3 ䷂ and Hexagram 4 ䷃ are upended images of each other. They appear on the chart with no box around them.

- **B–4 pairs of hexagrams balance each other's polarity**

In the chart's legend, these are the **B–4 Box Pairs**. These 4 pairs of hexagrams will balance out each other's polarity. In each pair, the first hexagram's lines are the polarized opposite of the second hexagram's lines. For instance, Hexagram 1 ䷀ and Hexagram 2 ䷁ are polar opposites…reversing images of each other. Indeed, they are so symmetrical that we cannot really tell if these two hexagrams also upend each other's image, as happened in **A**.

These 4 pairs of polarity-balancing hexagrams appear on the chart in undivided green boxes.

- **C–4 pairs of hexagrams flip-flop & balance each other's polarity**

In the chart's legend, these are the **C–4 Split Box Pairs**. The 4 **Split Box Pairs** both upend each other's polarity and also balance each out other's polarity. Thus the first partner's lines are the polarized opposite of the second partner's lines...yet the first partner clearly also upends or flip-flops to stand on its head to create the second partner! For instance, Hexagram 63 ☵☲ and Hexagram 64 ☲☵ both flip-flop and also balance out each other's polarity.

Only these 4 pairs of hexagrams on the chart very obviously use *both* options—both upending *and* balancing out. These 4 pairs of upending, polarity-balancing hexagrams appear on the chart in the split, rainbow-hued boxes.

7. Some philosophical symmetries

The hexagram pairs in King Wen order show an evident mathematical pairing according to line structure. However, those hexagram pairs are not organized primarily by the math's structural symmetries.

Instead, the hexagrams sequence by pairs along the rows in a philosophical flow that moves from conception and birth, through maturation and old age, to death and renewal. The hexagram twosomes along the way allow their paired dynamics to comment philosophically on each other, reinforce each other, or rebalance each other.

Each hexagram pair exhibits a reciprocity of attitudes and an entrancing symmetry of related meanings. The two hexagrams in each pair offer various mirror-twist reflections on each other's philosophical directives.

For example, Hexagram 49 ☱☲ *Revolution* boldly and abruptly overturns the established order. But its partner that is flipped upside down, Hexagram 50 ☲☴ *Cauldron,* is likened to a stew simmering in a pot, slowly transforming its hodgepodge of raw ingredients into a digestible, palatable meal. Thus, the hexagram of abrupt *Revolution* insists on immediate change, while the hexagram of *Cauldron* has an unhurried pace that changes multiple things gradually in a slow, evolutionary process.

Even more interesting to me is the fact the progression of all 64 hexagrams sets up like a storyboard. The tale begins when boy meets girl—when masculine Hexagram 1 *Assertive Heaven* meets feminine Hexagram 2 *Receptive Earth*. Together they generate offspring in the difficulty of birth that is described by Hexagram 3 *Laboring Birth*.

That arduous beginning, however, is supremely successful, for it launches the newborn (person, behavior, or idea) into Hexagram 4 *Learning*. Here the novice learner faces a daunting task: climbing the high mountain of knowledge that one encounters during a human life.

Hexagram 5 *Waiting* teaches a child enough patience to delay gratification, physical or mental. Then Hexagram 6 *Conflict* acknowledges the inner and outer conflicts of personal struggle. Hexagram 7 *The Army* enlarges the issue of conflict by rallying an unseen defense system that is embedded within society and in nature.

Long ago, the Zhou people lived in a rural fiefdom whose defense system was a citizen army embedded within the populace itself, like a reliable underground water table hidden within the earth, available to draw upon in time of need. This dynamic is depicted by the trigrams of *Earth* ☷ over *Water* ☵.

Hexagram 8 *Holding Together* shows a growing youngster how to keep one's wits together in time of stress by employing sublime values, a constant heart, and firm effort. Hexagram 9 *Taming Power of the Small* tests this new-found resolve when seemingly weak restraints nevertheless hinder reaching a goal, much as a heavy cloud cover promises rain to farmers in a time of drought, even though it may result in no moisture. Hexagram 10 *Treading on the Tail of the Tiger* (yet not getting bitten) teaches a youngster to meet unexpected threats or dangers yet not get bitten in the jungle of life.

The storyline moves on through all 64 hexagrams of the *Zhou Yi* order. Instead of extolling the values of external wealth and status, it cultivates the inner life meant to find and follow the dynamic flow of the Tao.

The storyline of the hexagram sequence charts the passage through childhood joys and tribulations on into adolescence, then into adulthood with its attendant desires, responsibilities, and challenges. For the I Ching, always the inner life is paramount. Learning to use the I Ching oracle to examine one's own motives and mental habits can open an awareness of the Tao's fractal patterns that hover beyond conscious logic.

After adulthood comes aging, which again brings new trials and perspectives. Hexagram 60 *Limitation* shows how to work with physical and mental limits. Hexagram 61 *Inner Truth* honors the core of inner truth that the Tao wants us to develop during a lifetime, with wisdom growing in us like a chick grows inside an egg. Hexagram 62 *Attention to Detail* counsels to use that inner wisdom to stay humble, respectful of divine order, and conscientious in small things, distinguishing between what is true and mere social noise.

The dynamic of Hexagram 63 *After Completion* brings things to an end, tied up in a tidy bundle, completed, with everything "in its proper place even in particulars," to quote Wilhelm/Baynes. But then comes Hexagram 64 *Before Completion* to overturn that finality by opening into something new, and thus the end becomes the start of something else.

The ancient Chinese text synopsized all this drama, hexagram by hexagram, into a concise set of rhyming couplets to describe the complete I Ching progression. For over a millennium, Chinese schoolchildren recited those rhyming verses to learn the philosophical gist of all 64 hexagrams—much as today's children know advertising jingles, hip-hop songs, or nursery rhymes.

The ancients knew that rhyme and rhythm possess mnemonic power. Verse is marvelously adept at getting past the left brain's logical limits by resonating the right brain into perceiving and honoring life's connective relationships.

But rhythm and rhyme can also disarm and disable your good sense. That's why a silly jingle from a TV ad can haunt you. It's why Hitler intoned his propaganda in a rhythmic incantation. It's why songs use rhythm and rhyme to growl out a protest against oppression, moan a plangent plea for love, or just coax you to buy.

Notice, those ancient rhyming verses in the 64 hexagrams of the I Ching did not coax their children to buy more stuff, but rather, to live more wisely.

Someday the Chinese will re-commence excavating the huge tomb of Chin Shi Huang and likely uncover many treasures. Archeologists are waiting to develop enough new technologies to do the excavation without destroying any aspects of those precious artifacts. They also want to ensure better safety standards that do not release any more toxic contamination from the tomb's hidden lakes and rivers of mercury used to replicate the waterways of an ancient world.

Who knows? Archeologists may even find the two long-lost I Ching books of *Lian Shan* and *Kuei Tsang*. What could they show us? I do not know. But it is fun to speculate on what may someday be found in that massive tomb. I even hope a finding may offer more clarity on the already-known hexagram orders. Archeology may someday even confirm that a deeper coding significance hides in the various orders of hexagrams.

In this series, we've already seen that the binary order of hexagrams can code for DNA 6-packs…that in turn hold a subcode for RNA codons…which can then be unpacked to deliver RNA message hexagrams that are already automatically sorted into their amino acid families.

Ancient China said the 64 hexagrams tap into the dynamics of universal body and mind flowing in the way of the Tao. Thousands of years later, Watson and Crick discovered the same mathematical paradigm in DNA that generates you and me, body and mind. From different epochs and angles, the two cultures found a polarized pair of pairs…that organized into polarized pairs of triplets…that bonded into 64 fractal patterns of co-chaos.

According to this TOE, both the genetic code and the I Ching math grow on the double p-tree embedded in nature's fundamental master code. Its co-chaos dynamics appear in many fractal variant subsets. These variants are seen in science as the broken symmetries of triplets or octets that appear at the macro, mid, and micro scales of our universe.

8. Using the I Ching oracle

Why does the I Ching oracle work? For indeed it does, if you can penetrate its layers of archaic vocabulary and feudal customs well enough to spot the dynamics. Each hexagram indicates the fractal form of its co-chaos pattern, but it does not fill in the specific details. Indeterminacy still exists. You determine its details by living out the choices that you make during that dynamic, including perhaps even opting for a better dynamic.

I suspect the I Ching works because its algorithm resonates with the master code that generated the universal body and mind, whose continual emergent flow includes us, body and mind. If we can grow aware of the cues that life is signaling, they can coax us into optimizing our own path in the flow of the Tao instead of floundering around in it. Heeding such cues will not make life perfect. Only easier. Kinder. Wiser. More aware. More meaningful. More appreciative. More able to solve problems.

Only your personal experience and well-kept records can show you whether the I Ching really works for you—or rather, whether you can learn its ways. Each hexagram answer gives an analogy for what is happening in your life. Its accuracy is not determined by blind belief, but rather, by your empirical judgment in assessing events over time.

Some humility is required for the logic-prone upon entering this analog, resonant realm. If you do not know the I Ching yet, I suggest that you try using it for three months. Religiously! (That was a joke.) No, just observantly.

Ponder your questions…and the oracle's answers. Look for a connective resonance, an emotional thread beyond the logical factors. Keep a record and check back occasionally on its answers to verify your memory, which can be remarkably faulty, skewed by a previous supposition, or perhaps resistant to the whole oracle notion.

To a skeptic, even when an answer seems apt, it can be readily dismissed as a quirky, weird coincidence. It is probably just chance, the ego stammers, or maybe a momentary suggestibility, or perhaps a mental blind spot. A fearful or defensive ego can label any answer that it receives as mere pointless garble.

You need to see the congruence of the I Ching's answers often enough over a long-enough time to realize how well it actually works. By keeping your own records and re-examining them at some distance from the heat of events and your own prejudicial blind spots, you may gradually discover that it works for you.

But you must also work for it. Before your ego can relax its defenses enough to let in the truth, you must want the truth enough to be honest with yourself…at least if you're like the skeptic that I used to be. If your logical, linear left brain does not shut down your holistic knowing, I suspect that over time, you will experience a connective thread of truth far beyond chance running through your I Ching answers. You may even find a wise friend here.

When you consult the I Ching, it offers you a hexagram's dynamic that resonates with you and the issue at hand. Using the dynamic of a co-chaos pattern and some metaphors, universal mind shows your own mind a dynamic that is relevant to your question. Yes, the I Ching is that direct, that relational. The fractal network underlying reality is just that thoroughly attuned to you. The Bible says that God's eye is on the sparrow. It can be a shock to realize that the universal mind's eye is on you. Each of us is always on its mind.

I must also warn you: there is some danger in pursuing the numinous. The closer you get to the source, the more you approach its grandeur. You may become lured, fascinated, besotted, inflated…or repulsed, frightened, confused, overpowered…lost in a rush of awe or awfulness. Remember, people can drown in the enchantment of the deep—whether it be the briny ocean, the deep well of black space, or the unplumbed depths of your own unconscious.

Make a special effort to stay clear-eyed. Recognize that the numinous can become a clarifying light or a blinding glare. Old myths tell us of mortals who looked upon Jehovah, Zeus, or other gods and were blinded, incinerated, or turned to stone. Those myths represent a psychological truth. The numen is awesome, and sometimes the truth of a revelation can feel just awful. So stay wakeful, considerate, and humble. Use the I Ching without turning glib, dismissive, or petrified by it…and most of all, without turning away from it.

Skeptic though I was at first, I investigated the I Ching, and slowly I changed my mind about it…and as you well know, changing your mind changes everything. It altered the events coalescing around me, and it brought me here to you.

Some would suppose that exploring lost books, lost codes, lost worlds, is too fanciful a waste of time, but it was mere speculation about a dream that led me to stumble upon the co-chaos paradigm. Somewhere at the fundament of universal being, the two realms of mind and matter really do merge. Seeking that source has delighted me. I hope it delights you, too.

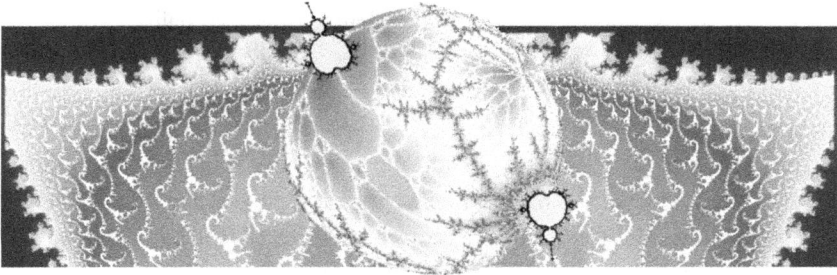

Chapter 14: Hexagram 3

1. See more with dreams

Some children are born of the body; some, of the mind. This series of books was born in my mind. I was notified of their birth in a dream on October 30, 1985, but when I woke up, I didn't realize what the dream meant. Not at all. Not until I brought that silly little dream of me birthing a litter of babies into the consulting room of Jungian analyst Elisabeth Ruf in Kusnacht, Switzerland.

I was laughing as I told her my short, ridiculous dream of multiple births, 6 of them. The first 3 had names, and the last 3 babies came tumbling out so easily—pop, pop, pop. But then I realized that she was staring at me.

It was not until I sat listening to her probe the dream's deeper meaning that I began to understand, slightly, what it portended. That dream is discussed in more detail in Volume 2, *Co-Chaos Patterns,* Chapter 8. And in Chapter 16 of my book on dreams, *Dream Mail,* is a follow-up dream on those newborns.

But here in Volume 3, I am merely discussing this birth dream specifically in relationship to Hexagram 3 *Laboring Birth.* Understanding the dynamic of Hexagram 3 has helped me write these books…although they did not pop out fast or easily…at least, not by my timetable. What a joy they became, however, and what a pain. Just like children. But without the counsel and caution of Elisabeth Ruf, I might have miscarried and not managed to bring forth even one book, much less this series that now includes 6 books.

As Frau Ruf and I talked, I told her that I'd awakened from the dream at exactly 5:00 AM. "Ah, the 5," she said. "It goes beyond the stability of the 4. You awoke to the creative ingenuity of the 5. It seeks and fathoms the esoteric."

I must've looked puzzled. I felt puzzled. After all, I'd walked into her office thinking this quick little dream of babies popping out like a litter of puppies was amusingly absurd. I'd already birthed two babies, so I knew it wasn't easy.

"I do not speak now of numerology," Frau Ruf said. "This is something else. Something deeper in the psyche. If you do not believe me, go ask Theo Abt (a Jungian analyst in Zurich). He studies the subliminal cues in numbers."

"Theo Abt," I said. "I'll note that name." (And I did. Eventually, I did analysis with Dr. Abt, and I also edited the English version of his book on numbers.)

Frau Ruf said, "The procreative power of the 5 is implicit in how 5 points can generate the star of Venus…and the pentagon of Mars. You can connect the points of each to find the sameness and the difference. This 5:00 AM awaking you had suggests that the masculine and feminine in you can come together now to nurture these dream children you've just birthed. Will birth into reality."

"Because I woke up at 5:00 AM exactly?"

"And because you noticed it. Synchronicity tells us things. If we listen." She went on to say that the dream was asking me to take care of something aborning, something new. She said these were not children of my body, but of my mind. "Creations of the mind. Things you must nourish with your mind."

2. Hexagram 3

Hexagram 3 is about the dynamic of giving birth. But do realize, birth happens in all sorts of ways. It brings forth all manner of children. If you insist on reading an I Ching hexagram as strictly literal, it may be quite difficult to find any real meaning in it for you…for instance, you might be a man who shrugs away the ridiculous notion that you could give birth, or even want to.

So approach your answer from different angles to realize what in it coheres for you. Remember to use both math and metaphor. The math comes from its hexagram structure made of two trigrams. They bond two fractals, two chaos patterns to create that hexagram's specific dynamic in the co-chaos system. But the metaphors come from a rural Chinese fiefdom over 3,000 years ago, using analogies and symbols that your own culture may not be attuned to.

Until you know a hexagram well, I suggest this: read several texts to interpret its meaning. While reading them, consider how the hexagram's dynamic applies to your situation. Utilize your left brain's linear processing power and your right brain's holographic pattern recognition to grasp its relevance to you…i.e., use both logic and emotion. Meditate on handling the hexagram's dynamic well in your own particular situation. Keep a record to test it and your responses.

Tapping into the I Ching offers a two-way conversation with the Tao. In each volume of this series, the last chapter explores I Ching interpretation by combining logical sequencing and analog examples.

This Volume 3 examines Hexagram 3, and the interpretation is my own. Below you'll see the hexagram number, its name in Chinese and English, and its hexagram, a mathematical figure. After that, you will find the *Image, Judgment, Hexagram Lines, Line Interpretation, Analogy, Analysis,* and *Example*.

First, read the *Image* and *Judgment* to get the basic dynamic of your hexagram.

Hexagram 3: 屯 *Laboring Birth* ≡≡ *Co-chaos Math*

Image

The shock of thunder cracks and rends the cloudy heights.

Birthing the future wracks the present to bring delights.

Helpers toil despite the stress to bow the rain.

Supreme success!

The Judgment

Difficulty in the beginning has great progress and success.

Benefit comes from being resolute and true.

Do not move without cause. Find helpers.

The Lines

Line 6
WATER — Line 5 ☆
Line 4
Line 3
THUNDER — Line 2
Line 1 ☆

Hexagram 3

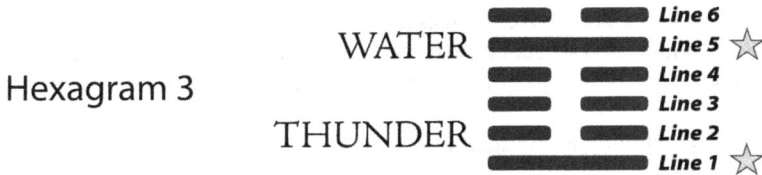

☆ **denotes the most important line(s)**—*usually 2 and/or 5.*

Now overview all the hexagram's numbered lines to understand the dynamic. *Line 1* sits at the bottom of the hexagram figure. Read upward to *Line 6* at the top. Next, apply any changing lines *in your answer to* add the nuances of their dynamics *to your own specific situation.* Add to that any insights you find in the *Image, Judgment, Interpretation, Analogy, Analysis,* and *Example.*

Line 1 —— ☆
Delay and hindrance block the birth.

Yang's strength and resolve show their worth.

Line 2 — —
Inept suitor? Yin will dismiss you;

She seeks late union and good issue.

Line 3 — —
Do not chase a goal into unknown land,

Desist for right now, or regret is at hand.

Line 4 — —
Blocked? Let it serve as a brief retreat

To win the strength of a good helpmeet.

Line 5 —— ☆
Yang's strength surges without discharge;

Wins small gains, but none are large.

Line 6 ▬ ▬
Frenzied gain despite yin fears,

Weeping streams of blood and tears.

3. The Line Interpretation

To interpret your hexagram answer, consider its basic co-chaos dynamic. Then flavor it with the specific influence of each changing line you received.

Line 1 ▬▬▬ ☆

In the lower trigram of *Thunder* ☳, yang is eager to act, but the way upward is impeded by the foreboding upper trigram of *Water* ☵. Storm clouds generate *Thunder's* shock. Advance is difficult. Despite a desire to act, this yang line must delay and seek help from other sources. Find a safe place to stay for now—literally "a boulder and a tree." The rock implies not moving, and the tree's safe place recalls the arterial tree that an unborn child sees as the "trunk" of the umbilical cord, the branching arteries that connect to the mother. The unborn is waiting safely now. You are counseled to bide your time safely, and the outcome will be advantageous.

Line 2 ▬ ▬

Line 2 is receptive yin, nudged by the yang line's vigorous push below it, exerting urgent pressure. But yin spies Line 5's yang in the trigram above. It is her true companion and correct suitor, so yin refuses that too-hurried alliance with the line below, postponing action. As a reward for this good judgment, Line 2 finally manages (after a hyperbolic 10 years in the ancient text) to join her bond-bigram, Line 5, and give birth. The analogy says birth's first yang contractions do not bring immediate delivery. A prudent course is to rest in yin pauses and choose the best time to act, not hurrying the procedure. Then the result is fruitful.

Line 3 ▬ ▬

Here is another, stronger warning to slow down. Rushing precipitously after a simple solution, a quick fix, the easy outcome, is like chasing a deer into unknown territory without a guide and getting lost in the wilds. Wisdom warns of hidden risks, so do not press the chase yet. Plunging forward now will only bring regret.

Line 4 ▬ ▬

Now forward momentum seems lapsed into retreat. This yin's energy is weak after those two yin lines just below it. Yet the yang line above is invigorating, with cycling animal energy that is reminiscent of muscle contractions (literally "horse riding in circles"). Yin falls back wisely, all the way back, even to companion yang Line 1, where she finds dependable strength. With yang above and below in steadfast, cycling support, yin can rest, assured that she will make progress. The idea here is to fall back to a rest position that becomes the basis for new forward motion.

Line 5 ▬▬▬ ☆

This is the most favored line in this hexagram. Yang struggles in powerful action, yet finds it difficult to tap the store of potential energy (literally "oil') to ease/grease the way. Why? Yang's vigorous movement toils in the center of the perilous trigram of *Water* ☵. Although yang action can propel the situation forward by small increments, it cannot do so with one great push.

Line 6 — —
 The dynamic of birth runs its inevitable course in many cycles of muscle contractions (as "riding the circling horse"). Retreat from the task is impossible. The water trigram bursts; blood and tears flow. For gentle yin, the stark reality is frightening, but it births a new life. This birth of new life feels fraught with terrifying uncertainly, but yin wisdom will accept life's cycling dynamics, thus easing the trauma already in progress to make this birthing of the new order a supreme success.

4. The Associations
Birth Pangs, Stressful Start, Nascent Struggle, Genesis, Emergent Life, New World Aborning, Inaugural Task, Resolute Beginning, The Straits of Passage, Your Own Association Regarding this Archetype

5. The Analogy
 Hexagram 1 *Assertive Heaven* and Hexagram 2 *Receptive Earth* come first in the I Ching sequence. They describe Father's yang assertiveness and Mother's yin receptivity. Interaction between them births something new. The co-chaos dynamic of Hexagram 3 ☳☵ *Laboring Birth* describes the labor of birth, using Hexagram 3's two trigrams to symbolize a thunderstorm in full turbulence. Dark storm clouds full of rain, the *Water* ☵ breaking, rides above *Thunder's* ☳ swift shocks of lightning. The rain, thunder, and lightning culminate finally in the sweet rainbow of success that signals the birth of something new.

 Laboring birth needs steadfast effort, so appoint helpers to bring the event to fruition. Active yang sits at bottom Line 1 of the hexagram and also in powerful Line 5. Between them sit the pauses of four yin lines, so yang energy must push onward to activate the birth. This hexagram's analogy of laboring to make a successful birth suggests that the turbulent flow of events is now moving beyond old ways to birth a new order in some area of your life.

 The new creation that emerges from this laborious birth wants to succeed in some form…physically, mentally, emotionally, or spiritually. It wants to establish a new generation, idea, attitude, or phase in the perpetual becoming that is life. Its difficult beginning has supreme success in birthing the new.

 Birth is a profound act, whether it be of a person, an idea, a business, or a sun. Here a potential gets materialized into reality. To bring forth a child, an idea, or a business requires labor. Hexagram 3's math sets its two trigrams into a co-chaos pattern indicating a strenuous passage, while its text counsels to ride the cycles of its momentum, both in pause and push, to fruition.

6. The Analysis
 The ancient Chinese name for Hexagram 3 ☳☵ meant *Bursting Forth* or *Sprouting Forth*. Birth is not easy. Delivery into a new order is fraught with trauma. There is no alternative, however, but to see the effort through. Its *Image*

describes the lightning and rain of a storm's stress that bring forth the glow of something new, much as a rainbow gleams in the world renewed after rain.

The rising progress within the 6 hexagram lines describes how to pace the labor wisely to get best advantage from every contraction, respite, and surge of events. In the lower trigram of shocking *Thunder*☳, Line 1 initiates the push for change, but despite a strong start, the way is difficult and advance is blocked. There is nothing to do about it now but sit snug and plan for the future. In Line 2, motion becomes even more difficult, for unruly instinct urges an immediate grab for the goal, yet reaching it is impossible right now. Wait for the appropriate time to surge ahead for success. Line 3 gives a dire warning to curb any fervent intention to chase after the fleeting goal into dangerous territory. Cool your impatience for now, or you'll regret it later.

Next, the upper trigram of *Water*☵ indicates that the only course now is to persevere on through the watery peril. (*Water* here symbolizes various human fluids, and in its linear canyon, the birth canal.) In Line 4, yin knows cycles of frenzied animal muscle are ahead, so rest is necessary now to resurge in later gains. Line 5 has yang power, but it cannot force the issue; only small gains are possible. In Line 6, the birth is at hand. Cycling contractions of animal energy push forth the result. Lashes of pain bring blood and tears. The paradoxical result of this painful labor is supreme success.

If you are engaged in this hexagram's dynamic, it can feel tiring, frustrating, and frightening...especially if you do not even understand that something new is aborning. Imagine giving birth without even realizing what is happening!

Scary indeed! Thus, if you receive this hexagram when you ask the oracle a question, it is heartening to realize that this co-chaos dynamic indicates something new is aborning. The difficult labor you experience is not futile; rather, it can generate a new, more evolved situation.

Cultures worldwide have told creation stories about the archetype of birth as a cosmic myth of profound significance. For instance, the book of *Genesis* centers around a tree in Eden. Psychologists speculate that the tree of Eden may honor at an unconscious level the prebirth period when an unborn baby can actually see the umbilical cord *in utero* if the mother stands in bright sunlight. The tree trunk is the umbilical cord; its branches are the mother's dark arteries and veins snaking around the fetus. An exit from the womb becomes an expulsion from that Garden of Eden to establish a new, independent life.

In summary, the difficult process of birth must move in a natural rhythm from first pangs to final delivery. Persevere and find helpers. Be content in the beginning merely to plan without taking overt action. Do not force things prematurely. Do not misjudge any respites in movement as setbacks. They are what allow you to pause and renew your energy. The big push will come, and even that, too, may at first seem ineffectual. Amid pain and tears of change, the tension finally turns into emergent ease. The result brings supreme success.

7. The Example

I have kept records with many examples of Hexagram 3, *Laboring Birth* occurring in my life, but one that springs to mind now about the I Ching is when I began to study calculus. Who, me?…study calculus? You must be joking!

How did that happen? In 1990–91, I became a visiting professor at Jinan University in Guangzhou, China, largely because it gave me access to study the I Ching. However, few educated Chinese at that time would even admit to having heard of it because it had become legal to use again only a year before.

At first, the only person who would talk about the I Ching with me was a scholar named Tan Shi-lin. His heritage was half-Chinese, half-German. He was an excellent and intriguing conversationalist on the I Ching, Chinese history, and Chinese culture. His gentle, beautiful wife Ye Sha also became a friend, and when I eventually left China, she gave me this delightful plate.

China Plate from Tan Ye Sha

The Tans soon introduced me discreetly to Zhang Luanling, a perceptive old professor of English, and a treasure hiding in plain sight there at the university. As a mentor, he elucidated the I Ching hexagrams very well, and he did the yarrow stalk procedure expertly. However, he was no mathematician able to discuss or deconstruct the co-chaos paradigm behind its algorithm's operations. I could tell it triggered a deep kind of knowing. But how?

After four futile months in early 1991, on the spur of the moment, I took a Saturday morning train to Hong Kong, then still under British sovereignty. I hoped a quick weekend in the city's pulsing momentum would jar me out of my impasse. All day I roamed Hong Kong, riding every sort of transportation I could find, from cab to ferry to double-decker bus to electric tram to Victoria Peak's funicular railway. Why? To gain momentum! I had wide access to people, libraries, and bookstores, but I traveled with no concrete idea of where to start.

Then on Sunday morning in my little room at rinky-dink Chungking Mansions ($100 Hong Kong a night/$12.50 US), I was brushing my hair as I turned on the black-and-white TV (extra fee of $8 Hong Kong/$1 US). Two men were discussing geometric progression. Calculus, the backdrop said. What!

I asked the I Ching if calculus or some other kind of math would help me deconstruct the I Ching algorithm. Its answer was unchanging Hexagram 3, *Laboring Birth*. Calculus or some other math? By noon, I was examining calculus books in Swindon's Bookstore to find something simple enough, yet thorough enough to take with me back to mainland China. To understand the math, I'd begun experiencing the trials of Hexagram 3, *Laboring Birth*.

Over time I realized that calculus was not the answer. Instead, I was facing a special kind of nonlinearity. So I bought more math books to feed my new infant body of knowledge on chaos theory. Eventually, I found the dp-tree's analinear math that generates co-chaos fractal patterns in events. Then in 2016, I found a paper explaining the algorithm that activates the I Ching query. It examined how the algorithm has three operators—the "intrication operator, turnover operator, and mutual operator"—as described in *I-Ching Divination Evolutionary Algorithm and its Convergence Analysis* by C. L. Philip Chen, Tong Zhang, Long Chen, and Sik Chung Tam.

8. Adele Aldridge: Hexagram 3, Line 3

Here is another view of Hexagram 3 *Laboring Birth* in the words and illustration by Adele Aldridge, a longtime explorer of the I Ching. The material below reflects her experience of Hexagram 3, Line 3.

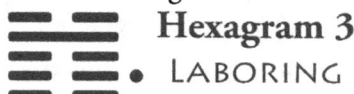

Hexagram 3
LABORING

Hexagram 3, Line 3

I cannot force events ahead of their time.
The seeds of my desire will mature in the future.

The history of the I Ching is one of the oldest in literature. The internet is its modern treasure trove of information. You can find free I Ching programs and thousands of years of relevant texts on the web. (The ebooks in this series provide links to many sites.) The I Ching is so old, so wise, so encompassing that it can embrace many approaches from different ages and cultures. I invite you to explore its historical depth and inclusive breadth without end.

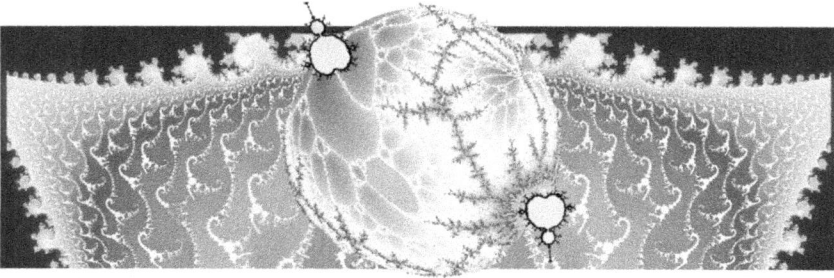

List of 20 Cosmological Questions

Earlier volumes in this series began with a discussion of 20 intriguing puzzles about the universe. Volume 3, *Tao of Life*, however, is mostly laid out to show you how I Ching hexagrams can be used to shorthand the genetic code, as well as describing some philosophical ideas. So in this third volume, I only list the 20 questions briefly to indicate the larger scope of the series.

In physics, many important questions still remain unanswered. This series examines 20 of these questions to see how they are answered in the Double Bubble TOE.

Question 1: What is our universe?

Question 2: What is the working shape of the universe?

Question 3: How did the universe begin?

Question 4: How deep in nature does fractal patterning go?

Question 5: Where did the original lost antimatter go?

Question 6: How did dimensions develop?

Question 7: What is gravitation?

Question 8: Why does the universe seem to expand constantly and ever-faster?

Question 9: Did the cosmic egg inflate in a hot Big Bang?

Question 10: Why do so many equations have important reciprocal solutions?

Question 11: Why is physics plagued with "impossible" infinities?

Question 12: What is electromagnetism? What is polarity?

Question 13: Why does light in a vacuum move at a constant speed? Not faster?

Question 14: Why won't Einstein's cosmological constant go away?

Question 15: Why can a particle-wave act like a particle OR a wave?

Question 16: How can two particles communicate faster than the speed of light?

Question 17: What is the neutrino?

Question 18: Is our universe designed to foster life?

Question 19: How will our universe end? Or will it?

Question 20: Is it possible to reconcile physics & philosophy in a TOE?

Series Summary

1. What is our universe?

This TOE says we live in the Double Bubble universe. Its two bubbles are conjoined, symbiotic mirror-twins with reciprocal properties of space, time, matter, and energy. Science sees our white-hole bubble above the *quantum* scale, where matter and energy emerge. It does not see a black-hole bubble conjoining our bubble at the far-tinier *mobic* scale where space and time emerge.

Known pole of gravity is in this bubble

1D space height

1D space depth

1D space width

1/2D time arrow

The White-hole Bubble

has SPACETIME…*its pole of receptive, negative* ━ *gravity attracts mattergy.*

3D space volume carries original matter as tardyon particles going up to the speed of light
&
The **½D time arrow** carries tardyon energy waves going up to the speed of light

Membrane interface at the mobic scale

Mobic Mirror *of* 2D space / 2D time ∞

The Black-hole Bubble

has TIMESPACE…*its pole of assertive, positive* ✚ *gravity repels antimattergy.*

1/2D space arrow

1D time past

1D time future

1D time present

Lost pole of gravity is in this bubble

The **½D space arrow** crush-converted original antimatter into tachyon particles of imaginary number mass going above the speed of light ^{2}
&
3D time volume carries tachyon energy waves going above the speed of light ^{2}

The Double Bubble universe has 11 dimensions

Our upper bubble has the *spacetime trident* of contiguous 3D space with a one-way, ½D arrow of time, plus one pole of gravitation and the original matter. Its 3D space holds many material 3D structures morphing on the arrow of time, but its particle-waves are slowed down to the speed of light.

The lower bubble has the *timespace trident* of 3D time with a one-way, ½D arrow of space, plus gravitation's "lost" pole, plus the original "lost" antimatter that was long ago crush-converted by that lower bubble's meager ½D space into tachyon particle-waves going above the speed of light[2]. It powers a huge, unified mind constellated in vast 3D energy patterns in the lower bubble's 3D time.

The mirror-twin bubbles conjoin at a membrane interface of ultra-tiny, mobic pores called *mactors*. Each pore combines traits of a Mobius band and a Lorenz attractor, hence the name of *mactor* for its dynamic at that ultra-tiny scale.

How did space and time begin? The cosmegg set a dimensionless point with an *on*-pulse of being. A second *on*-pulse sketched a 1DD line of polarized tension with two poles: space and time. A third *on*-pulse set a 2DD triangle with two polarized faces: 2D space and 2D time. Polarized tension ran around both faces on a χ path much like an infinity ∞-loop.

Just one more *on*-pulse turned that 2DD-triangle into a 3DD tetrahedron. It had two volumes: 3D space and 3D time. The outer volume of 3D space projected far above the mobic scale, while the inner volume of 3D time projected far below it. Together they made an hourglass cell. That single cell replicated many times. All cells merged into our holographic bubble of 3D space above the mobic scale; below it, into a holographic bubble of 3D time.

The Double Bubble hologram merged all its ∞-loops and projected them, so now they 8-loop across both bubbles, switching polarity as they cross the interface, creating the tensor network of a single yet ubiquitous dimension made of two ½D arrows moving on an endless, polarized 8-loop across both bubbles.

The upper and lower halves of this 8-looping tensor network are polarized per bubble as either ½D time or ½D space. We have the time pole. We experience it as the point of constant *now* moving forward on the arrow of time. But that other bubble has the constant point of *here* moving backward on its arrow of space. *Here* is the only location possible in that other bubble. (There is no *there* there.)

2. Count the dimensions of this kleiniverse

Our universe has how many dimensions? Count 3.5 dimensions per bubble, making 7 dimensions in both bubbles. Count 4 more dimensions at the mobic scale itself, where every mactor's mobic warp generates 2DD triangles with polarized faces of 2D space and 2D time. This totals 11 dimensions in a layout of complementary space and time that is symmetrical across both bubbles.

The dynamic of the Double Bubble recalls a Lorenz attractor. Its two domains are the upper and lower bubbles. Its 3D space and 3D time act as three coupled Ordinary Differential Equations (ODEs) iterating along the arrows of ½D time and ½D space to evolve the nonlinear solution of reality emerging in both bubbles. The reciprocal laws of physics and the reciprocal scaling of space and time, matter and energy let both bubbles fit inside each other as a *kleiniverse*.

3. The master code uses four primals

This TOE says our universe is a huge, living organism whose fractal structure is generated by a co-chaos paradigm that iterates in self-similar patterns on many scales. Its master code uses four primals: space, time, matter, and energy. This polarized pair of pairs sort into two *carriers*: space and time...and two *cargoes*: matter and energy, polarized such that space carries matter, and time carries energy.

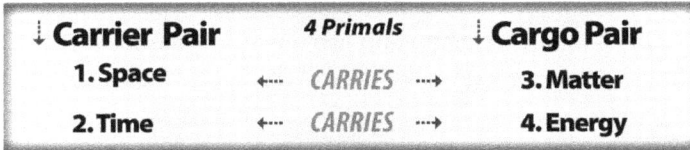

↓ Carrier Pair	4 Primals	↓ Cargo Pair
1. Space	←··· *CARRIES* ···→	**3. Matter**
2. Time	←··· *CARRIES* ···→	**4. Energy**

Panel: the 4 primals are a polarized pair of pairs

Our 3D space bubble has tardyonic particle-waves in vast material structures of self-similar, evolving 3D patterns. The 3D time bubble has speedy tachyonic particle-waves in vast energetic constellations of self-similar, evolving 3D patterns. In both bubbles, intricate detailing on many scales recalls the Mandelbrot set.

Both bubbles cooperate to refresh their space-time forms and update their matter-energy cargoes at a rate that makes our holographic universe appear to be smoothly continuous to our senses and mechanical tools above the quantum scale, the smallest scale known to current physics. In the universal body, old configurations decay and new ones develop. We tiny organisms in the upper bubble experience this flux as the emergent events of ongoing reality.

4. The genetic code is a variant of the master code

How can we decipher the master code that iterates the universe? We can study a lesser variant, the familiar genetic code that iterates us. It offer us some clues.

CLUE: DNA uses four base molecules: **T**hymine, **C**ytosine, **A**denine, and **G**uanine. Its polarized pair of pairs sort into two *pyrimidines*: **T** and **C**—and two *purines*: **A** and **G**. They are polarized such that **T** bonds with **A**, and **C** bonds with **G**.

CLUE: The four base molecules can pair-bond by triplets (*codons*) to make 8 × 8 = 64 molecular 6-packs on the double helix to iterate and maintain the bodies of all evolving species. Old configurations of individual organisms decay, and new ones develop. We experience this flux as the emergent lives of ongoing species.

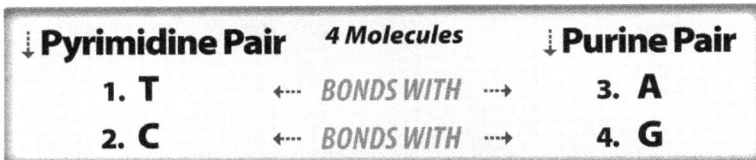

↓ Pyrimidine Pair	4 Molecules	↓ Purine Pair
1. T	←··· *BONDS WITH* ···→	**3. A**
2. C	←··· *BONDS WITH* ···→	**4. G**

Panel: the 4 DNA molecules are a polarized pair of pairs

5. Our Rosetta Stone: I Ching, genetic code, master code!

This TOE says the ancient I Ching hexagrams of China offer a math shorthand for this paradigm. It grows on a bifurcation tree that is both doubled and polarized, i.e., it is a *dp-tree*. The I Ching's easy math can shorthand the genetic code, and their kinship gives us a Rosetta Stone with three scripts, two known and one unknown. The two known codes can help us decipher the unknown master code.

CLUE: Like DNA, I Ching math figures are also a polarized pair of pairs. They sort into two stable bigrams and two unstable bigrams: stable yin ☷ , stable yang ☰ , changing yin ⚎ , and changing yang ⚍ .

Yin-based ↓	4 Bigrams	Yang-based ↓
stable yin **1.** ☷	←⋯ *STABLE PAIR* ⋯→	stable yang **3.** ☰
changing yin **2.** ⚎	←⋯ *CHANGING PAIR* ⋯→	changing yang **4.** ⚍

Panel: the 4 bigrams are a polarized pair of pairs

CLUE: The I Ching math develops on a dp-tree, and it can shorthand the other two code variants. Below, the dp-tree has three levels of polarized forking above and below a neutral 0 seed in the middle. On the dp-tree, *minus* ‒ stands for a yin — — fork. *Plus* + stands for a yang —— fork. Its first level of forking outward develops ‒ *yin* and + *yang* poles. The second level outward develops the four *bigrams*. The third level outward develops eight *trigrams* above and below.

Each trigram is a *vertical* period 3 window (*vp3*) defining a chaos pattern. This postulates a vital variant on Yorke and Li's *horizontal* period 3 window (*hp3*) that defines a chaos pattern in their seminal paper *Period Three Implies Chaos* (1975). In each vp3, addition by 2s, period-doubling, and exponential power together create a nonlinear chaos process so special that I call it *analinear*.

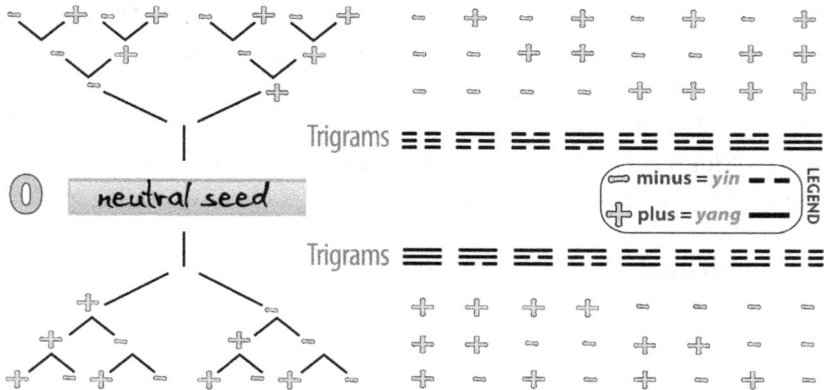

The dp-tree has 8 × 8 polarized vp3s = 64 co-chaos patterns

CLUE: Each trigram's math describes a *chaos pattern*. The dp-tree can pair-bond its trigrams into 8 × 8 = 64 *hexagram* 6-packs of *co-chaos patterns*.

Our Rosetta Stone's triple play features the familiar genetic code, the ancient I Ching, and an unknown master code. The first two codes have some shared traits that will help us discern features of the master code at the mobic scale, where polarized pulses organize into triplets of information in myriad mactors. The triplets then pair-bond into 8 × 8 = 64 co-chaos patterns that develop our universe's emergent properties. They project the Double Bubble's space-time skeleton and flesh out its matter-energy body to evolve its huge, ongoing life.

6. What are we?

This TOE says our Double Bubble universe lives, and we are like microbes living in its gut, oblivious to its larger aims. In our white-hole bubble of 3D space, we tiny, diverse, walkabout minds are powered by particle-waves of slow tardyon energy. But the black-hole bubble of 3D time holds a single, giant, unified mind that is powered by zippy tachyon energy moving at more than lightspeed[2].

Many tiny bodies with portable minds inhabit the upper bubble. But when a mind is released from ego identification by sleep, trance, meditation, or other means, it can tap into aspects of that unified mind in the tachyonic cloudbank of 3D time (some call it God or Mother Nature) processing the data of concurrent past, present, and future. That unified mind evolves its huge, beautiful, and diversified universal body. For instance, in the upper bubble, it established the far-flung galaxies and tiny micro-organisms under rocks that hold our attention.

The universal mind even delivers dreams to us nightly, in dramas that address our specific needs, fears, and hopes…but most of us have forgotten how to translate its symbolic lingo. Relearning it can cultivate a sixth sense, an ability to tap into nature's basic patterns by deep-see diving. We can access info in the tachyonic cloud of the lower bubble via shared intention and resonance. It recognizes and responds to whatever is in you—so go carefully and with good intentions. Treated wisely, it can heal and unify us, body and soul, layer by layer.

The minds in both bubbles contribute in various ways to the thrust of universal existence, which has a greater purpose beyond our own human preoccupations. Our universe plans a wider future for us as we become more conscious of our place in the whole. It has been patiently cultivating its universal life, including us among its myriad forms, hoping to evolve us enough to recognize that it too lives…and further, coaxing us to divine that there is something even greater beyond. We have the chance to acknowledge, share, and improve this destiny.

Blurb and Reviews

The Tao of Life
THE FRACTAL GIFT

Katya Walter, PhD

BIOGRAPHY

Katya Walter has a Ph.D. with an interdisciplinary emphasis from the University of Texas at Austin. She spent 5 years of post-doctoral study at the Jung Institute of Zurich, and a year of post-doctoral study in China. Dr. Walter taught in colleges and universities in the USA and abroad for 16 years before focusing on writing and lecturing. She has given numerous workshops on the I Ching, chaos theory, synchronicity, and dreams in the United States and Europe.

-:=:-

FROM THE EDITOR

Is this science fiction? Some say so, but it's not. Science and mysticism merge in this stunning new paradigm. It spans DNA, gravitation, chaos theory, dark matter, and remote viewing!

Western science explores cosmology. Ancient China's I Ching follows the Tao. They merge in a master code that generates our fractal universe. This series reveals that master code.

This book is Volume 3 in the dazzling *Touching God's TOE* series, 4th edition. In this volume, Katya Walter, Ph.D., shows how correlating the genetic code with ancient China's I Ching provides a Rosetta Stone to decode the master code that generated our universe. She also explores some of its philosophical aspects, including a huge, unified mind that exists in nature itself, accessible via dreams, remote viewing, and other altered states.

This series began as one volume, *Chaosforschung*, published in German in 1992, then as *Tao of Chaos* published in English in 1994. That book was later split and amplified into Volumes 2 and 3 of this series, *Touching God's TOE*.

Volume 3 has 14 chapters in 112 sections. It includes a *Series Summary*, *Bibliography*, and *Reviews*, along with 104 listed images, graphics, and charts. The color ebook version has an interactive table of contents and 86 e-links that act as informative footnotes. Its text is completely searchable and receives electronic updates. It is also hand-edited to hold color graphics that allow greater distinctions in images and charts. Consider getting both the print and ebook versions of this book for a greater range of information and versatility.

-:=:-

PRAISE FOR THE *TOUCHING GOD'S TOE SERIES*

What an interesting and inspiring writer… interesting scientifically and inspiring metaphysically! I have traveled widely, but never on a roller coaster of dimensionality before! It makes Flatland look—well, flat. Quantum organics reveals how space, time, matter, and energy mirror aspects of our DNA. And the author's take on what animals think is shockingly possible! You'll never regret picking up this series and reading it. It will take your mind to new places, and it will lift your soul along the way.

Lynn Hayden
Consultant, Singapore Institute of Management

"I find the *Double Bubble Universe* the most promising of all the TOEs being proposed currently. It involves a new topological model spanning all levels of reality and "deep-see diving" into fractal pattern recognition. It answers far-reaching questions such as 'How did our universe begin?' and 'How are telepathy and remote viewing possible?' This model deserves careful reading by the best minds of our time."

Oliver Markley, Ph.D.
Professor Emeritus, Human Sciences & Future Studies

"Are you smarter than a fifth grader?" Better yet, can you bring the clarity of a child's fresh perspective to a Theory of Everything (TOE) that reinterprets standard physics data to reveal a stunningly new and elegantly symmetrical model? If you can, then this book's for you.

Dr. Katya Walter shakes the foundations of currently accepted concepts about physics, metaphysics, and the nature of consciousness. She offers a comprehensive exploration of the way fractal chaos theory forms the underlying structural dynamic that creates and allows for the ongoing evolution of both mind and matter. She also addresses the relationship of mind and matter - physically, spiritually, and philosophically - in ways not previously presented elsewhere.

This seems like a good place to mention that no weeping, wailing and gnashing of teeth are required when reading this book… it is well written in a way that is comprehensible to a general audience as well as for scientists.

If you can set aside any skepticism and/or preconceived notions long enough to allow lucid consideration of the concepts she proffers here, you may be the first on your block to recognize her Theory of Everything as the dawn of the brightest new paradigm since Newton.

Brenda Kennedy
Reader

I can't decide if this is fact, sci-fi, or psi-phy. Whatever, it is truly fascinating. A brain gym of possibilities!

Frank Patterson
Aerospace Engineer

My guess is that your natural reader would be a non-scientist who wants to put science and philosophy together in a coherent mental image.

David Booth, Ph.D.
Mathematician, Inventor
-=:=-
I cannot emphasize enough how much I love this book. It makes the most current information about quantum physics into a conversation that can span the thinking styles of both scientists and spiritists. Katya is a dedicated dreamer, and a receiver of concrete knowledge in frontier quantum physics. There should be no separation of physics and metaphysics. There should be fluency and grace and relation to both subjects. This book achieves an understandable explanation of our human experience of dimensionality, and of our fractal nature. She proposes a new Theory of Everything (T.O.E.) If you want the most original elegant synopsis of our existence, which uncovers the mysterious forces of nature (including gravity) and therefore our consciousness, buy this book. You will have an "Aha" moment, and then you will be with me, saying, "Every life-student should be so lucky to have been exposed to Katya Walter. Reading *Double Bubble Universe* is like being in Einstein's living room."

Jennifer J. Colbert
Reader
-=:=-
Simply brilliant! And I mean that literally. The clearest explanations are the least complex, and Dr. Walter has managed to take ideas from advanced physics, express them simply, then turn around and analyze the physics to present a clear, simple, and straightforward new paradigm for how the universe works. This is the simplest physics book I've ever read, because of Walter's brilliant use of language that makes these complex concepts entirely understandable. The interweaving of her 'journey-to-the-aha' adds a profound metaphysical understanding of how our universe works from the inside out. You won't regret buying this book.

Anne Beversdorf
Reader
-=:=-
The author of this extraordinary book has a rare combination of qualities: an astonishing depth of vision and a genuine modesty. *Double Bubble Universe*... is exploring Katya Walter's theory of everything (TOE). A TOE is the Holy Grail of modern physics. A theory that reconciles the billiard-ball predictability of Newton's Laws with the mysterious goings-on at the Quantum level.

Dr. Walter's book proposes that Physics is blind to another domain in the universe which she describes clearly and patiently with easy-to-grasp imagery... the book really gets you thinking. In a book of this scope, it's very refreshing to find that the author has a gentle, conversational style and an open-minded approach towards the reader.

For example, she writes "Consider this a journey into possibility. I don't mind if you treat this as science fiction, science fact, an amusing tale... or purely just diverting balderdash... take it as you will and let it take you where it will."

... where it leads is to the I Ching, the ancient Chinese oracle that, according to Katya Walter, has: "unique fractal shorthand in a coded way that can merge physics and metaphysics."

Extensively referenced and full of diagrams—I really enjoyed this book and I'm looking forward to the next one in the series.

Mick Frankel
I Ching consultant-London, UK

-:::-

"This is the best book on this topic I've read and I've read a lot of them. Solid research on the science end without claiming unproven conclusions. The author simply explains her own TOE which she presents in a logical easy to understand manner. I appreciate her ability to speak to both the spiritual and scientific audience. Very thought provoking.

Winnie Hiller
Reader

-:::-

I think Katya Walter is a genius in that she can translate her right brain insights into left brain analysis with striking correlations and patient explanations. In this book, she's drawing on all her others to outline a sort of unifying theory of everything. Her discovery of the primordial pattern embedded in every level of creation, the "Master Code," is as significant as it sounds. Drawing on her first book, "The Tao of Chaos," she explains that this fractal pattern is originally created by the two primal pairs of opposites: space and time; matter and energy. She then follows the natural implications of that pattern to assert that there is a mirror opposite universe to ours of one-half dimensional space and three dimensional time. In her theory the missing or hidden parts of the pattern that we observe in physics (for example, the "arrow of time") are found in that mirror universe that she calls the Double Bubble universe."

Yeah, the "theory of everything" is a big assertion. Katya Walter's ideas are brave and bold- and impossible to prove. But, as a metaphysicist, she can't wait for the astro and nuclear physicists to catch up. Her books are sort of a field guide to physical reality for modern-day mystics. She explains her ideas through the models of biochemistry, a little math, geometry, and what she calls "the shorthand of the I Ching." She also includes her personal thoughts and dreams with her careful explanations of mathematics and physics. I'll admit, the mix takes getting used to. Yes, it's weird- but worth it!

Like any genius, the author is unconventional and eccentric, which could cause some people not to take her seriously. That would be a mistake, as a careful reading reveals an extremely intelligent and logical woman who asks the right questions. She simply doesn't stop asking, and may go a lot farther than most people are comfortable with, given that we may never have scientific proof for any of this. But, in this era of string theory which proposes many additional unknown dimensions, I wish the physicists would read her books. She could point them in the right direction, and may even save them some time with her simple and elegant theory of everything.

Erin Rose
Yoga Teacher & holistic Health Therapist

-:::-

Dr. Katya Walter's book *Double Bubble Universe* unites cutting-edge scientific research with her own inner 'deep see diving'. She accomplishes the incredible feat of inciting a paradigm-shift in the reader (to the realization that a love-intelligence underlies and pervades the physical universe), bringing her TOE to life! (unlike any other TOE I've read). Quantum mechanics, a physics of cold dead space, births 'quantum organics,' a science of the fractal aliveness of the universe!

Dr. Walter skillfully creates an enjoyable, light read—provocative, funny, and digestible, dealing with perhaps the 'heaviest' topic of all—the structure and meaning of the creation and evolution of the universe. Read this book to witness the wedding of science and heart—watch how every whirling particle spins in the same wind as love's art! Who knows what could bubble up?!?

Peter Craig
Licensed Professional Counselor

I highly recommend this book for those who believe the current scientific paradigm is incomplete, and are looking for new explanations to fill the gaps. The *Double Bubble Universe* is a space-time, matter-energy, symmetry explanation of the physical universe. It introduces a scientific explanation of the physical laws of the universe with 20 questions. Written in layman's terms, Katya Walter's book encompasses the melding of Science and Metaphysics in which she intersperses and interweaves a personal dream with frontier science. Katya's writing skills are extensive and second to none, coining phrases that are truly inspired and unique.

Don Switlick
Institute for Neuroscience & Consciousness Studies

Replete with deep scientific insights that answer previously unanswerable questions yet accessible to lay readers, Ms. Walter's book offers the most comprehensive and useful T.O.E. ever. Comprehensive in that it not only explains reality from subatomic levels to the most macro perspectives, but it also links the physical and the spiritual and connects the evolution of the universe to the evolution of consciousness. Useful in that its elegant explanation of physical reality has implications that naturally lead one to contemplate how to live one's life more effortlessly and authentically. I highly recommend this book to all those truly thirsting to understand everything.

Kevin Blackwell
Stocks & Bonds Analyst

Have you ever wondered why we (human species), considering that which most concur is ineffable, continue effing it up? I've always thought having minds dead set on figuring things out in combination with phenomenon that exceed our ability to do so could be called 'God's dirty trick.' Dr. Walter has taken just such issues and playfully made a case worthy of consideration while mercifully maintaining the topic's ultimate ineffability. I found myself intellectually giggling throughout this read. Her consideration of the I Ching and our DNA alone is worth the purchase of her books.

MIL
Reader

I have read several other TOE books…the latest being Tom Campbell's My Big TOE. Wanted to read this one and see what new perspectives were explored. Was pleased to find that BOTH books stand on their own, and each adds new information without contradicting the other!! (So this is a "Must Read"!)

Great visual-inducing analogies and metaphors. Enjoyed reading about her personal background, leading to development of this book.

Descriptive down-to-earth language, even tho, on occasion, I have to look up a word in the dictionary!…which means you learn new concepts and words as you go…. The subject matter covers questions we all ask at one time or another…and the answers are creative and original…makes you think and gives perspective.

Hyphenated, descriptive words are used where needed, supporting the requirement for "hyphenated sciences" and new words to explain some of the more ephemeral aspects of mind and consciousness. Very insightful, creative application of recently-discovered fractal phenomena to explain its basic principles to everything in the universe. Plenty of references and web sites for the reader who wishes to explore further.

James Beal, Ph.D.
Aerospace & Electrical Engineer

Katya Walter's series starting with the *Double Bubble Universe* integrates immense questions and insightful answers about the cosmos. She uses data, analogies, graphs, images, and stories that resolve together into one bubbling statement. A must read….

Rowena Pattee Kryder, Ph.D.
Dynamics & Foundations of Co-Creation

Katya Walter is that rare writer who can merge so-called opposing systems, like science and metaphysics. For me, with a PhD in Literature and Communication, and a serious teacher of A Course In Miracles, and having had spiritual experiences myself, I am so grateful that she brings it all together, so I no longer have to wonder if I should trust those marvelous "intuitive" experiences enough to share them with others, without fear of ridicule. Just read Katya Walter if you think this is not possible. Thanks, Katya. We need you.

Helen Bonner, Ph.D.
Author

Note from the Author

I FIND MANY MORE PEOPLE READ THIS BOOK THAN BOTHER TO REVIEW IT.
IF THIS BOOK WAS INTERESTING TO YOU, PLEASE REVIEW AND RATE IT.

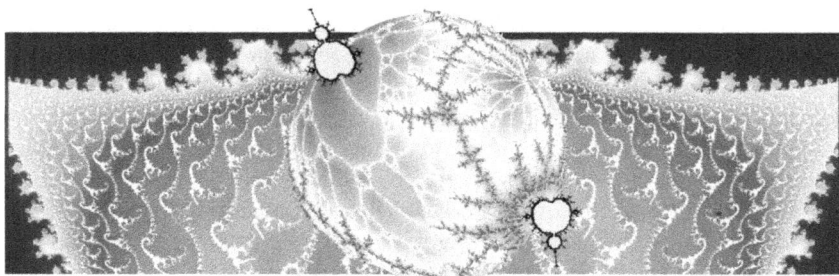

Bibliography

This is by no means all the books I consulted on writing this series, but these seemed to me the most relevant to this volume.

Anthony, Carol. *A Guide to the I Ching*. Anthony Publishing Company. 1980.

Barrett, Hilary. *Ching: Walking Your Path, Creating Your Future*. London: Arcturus Publishing. 2010.

Barrow, John D., with Frank Tipler. *The Anthropic Cosmological Principle*. Oxford: Oxford University Press. 1988.

Benson, Frank, editor. *The Dual Brain: Hemispheric Specialization in Humans*. New York: The Guilford Press. 1985.

Borek, Ernest. *The Code of Life*. New York: Columbia University Press. 1969.

Bowers, Kenneth, & Meichenbaum, Donald. *The Unconscious Reconsidered*. New York: John Wiley & Sons. 1984.

Buchanan, Keith, with Charles P. FitzGerald & Colin A. Ronan. *China: The History, the Art, and the Science*. New York: Crown Publishers. 1981.

Butler, Christopher. *Number Symbolism*. London: Routledge & Kegan Paul. 1970.

Campbell, Joseph. *The Masks of God*. New York: Viking Press. 1964.

Capra, Fritjof. *The Tao of Physics*. New York: Bantam Books, 1977.

Carus, Paul. *Chinese Astrology: Early Chinese Occultism*. La Salle, Ill.: Open Court. 1974.

Cole, K.C. *Sympathetic Vibrations: Reflections on Physics as a Way of Life*. New York: Bantam. 1984.

Cook, Norman D. *The Brain Code*. London: Methuen. 1986.

Coward, Harold. *Jung and Eastern Thought*. New York: State University of New York Press. 1985.

Culling, Louis. *The Incredible I Ching*. New York: Samuel Weiser. 1969.

Da Liu. *I Ching Numerology: Plum Blossom Numerology*. New York: Harper & Row. 1950.

Davies, Paul. *The Cosmic Blueprint*. London: Unwin Paperbacks. 1989.

Davis, Philip & Hersh, Reuben. *The Mathematical Experience*. Boston: Houghton Mifflin. 1981.

De Bary, William Theodore; Wing-tsit Chan; Watson, Burton. *Sources of Chinese Tradition*, Vol. 1 & 2. New York: Columbia University Press. 1960.

Dewdney, A.K. "Computer Recreations." Scientific American, August 1985.

Doczi, György. *The Power of Limits*. Boulder: Shambhala Publications. 1981.

Ellenberger, H.F. *The Discovery of the Unconscious*. New York: Basic Books. 1970.

Fairbank, John. *The Great Chinese Revolution: 1800 to 1985*. New York: Harper & Row. 1986.

Fiedeler, Frank. *Die Wende: Ansatz einer genetischen Anthropologie nach dem System des I-ching*. Berlin: Kristkeitz 1977.
   Yin und Yang: Das kosmische Grundmuster in den Kulturformen Chinas. Cologne: DuMont, 1993.

French, A.P. *Vibrations and Waves*. New York: Norton, 1971.

Fromm, Erich. *The Forgotten Language; An Introduction to the Understanding of Dreams, Fairy Tales and Myths*. New York: Grove Press, 1951.

Fung Yu-lan. *A History of Chinese Philosophy*. Trans. Derk Bodde.: Princeton University Press. 1952.

Gao Heng. Article translated from *Wenshizhe* by Edward L. Shaughnessy. Zhouyi *Network Newsletter*, No. 1, Bowdoin College, Brunswick, Maine, January 1986.

Gardner, Howard. *Art, Mind and Brain: A Cognitive Approach to Creativity*. New York: Basic Books, Inc. 1982.

Gardner, Martin. "Mathematical Games" in *Scientific American*, January, 1974.

Gascuel, Olivier and Danchin, Antoine. *Data Analysis Using a Learning Program*, Paper. 1986.

Gatlin, Lila. *Information Theory and the Living System*. New York: Columbia University Press. 1972.

Glass, Leon & Mackey, Michael. *From Clocks to Chaos: the Rhythms of Life*. Princeton: Princeton University Press. 1988.

Gleick, James. *Chaos: Making a New Science*. New York: Viking. 1987.

Gordon, Rosemary. "Reflections on Jung's Concept of Synchronicity," *Harvest*, Vol. 8, pp. 77-98. 1962.

Hayes, Brian. "The Invention of the Genetic Code" in *American Scientist*, January-February 1998, Vol. 86, Number 1, Page. 8

Heisenberg, Werner. "Scientific and Religious Truths." Seen as a typed essay.

Hofstadter, Douglas. *Metamagical Themas: Questing for the Essence of Mind and Pattern*. New York: Basic Books, Inc. 1985.

Hook, Diana Ffarington. *The I Ching and Mankind*. London: Routledge & Kegan Paul. 1971.

Horowitz, Seth. *The Universal Sense: How Hearing Shapes the Mind*. New York: Bloomsbury USA. 2012.

Hughes, E. R., trans. & editor. *Chinese Philosophy in Classical Times*. New York: Dutton & Co. 1942.

Jaynes, Julian. *Origin of Consciousness in the Breakdown of the Bicameral Mind*. Boston: Houghton Mifflin. 1976.

Jenkins, R.C. *The Jesuits in China*. London. 1894.

Jin, Guantao, Fan Dainian, Fan Hongye, & Liu Qingfeng. "The Evolution of Chinese Science and Technology" in *Time, Science, and Society in China and the West*. Volume V of *The Study of Time*. Edited by Fraser, J.T., Lawrence, N.,

& Haber, F.C. Amherst: University of Massachusetts Press. 1986.

Josey, Alden. "Molecules as Mandalas." Unpublished essay.

Jung, C.G. "Four Lectures on the Chakra Symbolism of Tantric Yoga and the Kundalini System" (1932). New York: *Spring Annuals of 1975 & 1976.*

Jung Young Lee. *Principles of Change: Understanding the I Ching.* Secaucus, New Jersey: 1971.

Kelso, J.A.S., Mandell, A.J., & Shlesinger, M.F. *Dynamic Patterns in Complex Systems—Conference Proceedings. Teaneck, New Jersey: World Scientific. 1988.*

Kramer, Johnathan D. "Temporal Linearity and Nonlinearity in Music." in *Time, Science, and Society in China and the West; The Study of Time,* Vol. V. J. T. Fraser, N. Lawrence, & F. C. Haber, editors. Amherst: University of Massachusetts Press. 1986.

Kreutzer, Carolin S. "Archetypes, Synchronicity and the Theory of Formative Causation," *Journal of Analytical Psychology,* Vol. 27, pp. 255-262. 1982.

Kuhn, Thomas. *The Copernican Revolution.* Cambridge, Maine: Harvard University Press. 1957.

Lattimore, Owen. *Studies in Asian Frontier History.* London. 1962.

Lepore, Franco; Ptito, Maurice; Jasper, Herbert, editors. *Two Hemispheres—One Brain: Functions of the Corpus Callosum.* New York: Alan R. Liss, Inc. 1984.

Legge, James, trans. *I Ching: Book of Changes.* Published in English in 1899.

Leibniz, G. Wilhelm. *Zwei Briefe über das Binare Zahlensystem und die chinesische Philosophe.* Munich: Belser. 1968.

Lex, Barbara. >See D'Aquili, Eugene G., Laughlin, C.D., & McManus, J. *The Spectrum of Ritual: A Biogenetic Structural Analysis.*

Liu Dajan. *A Preliminary Investigation of the Silk Manuscript Yjing.* Trans. from *Wenshizhe* by Edward L. Shaughnessy. Shandong: Shandong University. 1985.

Lovelock, James. *Gaia: A new look at life on earth.* Oxford: Oxford University Press. 1979.

Mandelbrot, Benoit. *The Fractal Geometry of Nature.* New York: Freeman, 1977.

Milner, Brenda, editor. *Hemispheric Specialization and Interaction.* Cambridge, Mass: MIT Press, 1974.

Moore, Steve. *The Trigrams of Han: Inner Structures of the I Ching.* Wellingborough, England: The Aquarian Press. 1989.

Morris, Eleanor B. *Functions and Models of Modern Biochemistry in the I Ching.* Taipei: Cheng Chung Book Company, 1978,

Nalimov, V.V. *Realms of the Unconscious: The Enchanted Frontier.* Philadelphia, Pa.: ISI Press. 1982.

Needham, Joseph. *Chinese Astronomy and the Jesuit Mission.* The China Society. 1958

  "Science and China's Influence on the World," in *The Legacy of China,* edited by Raymond Dawson. Oxford: Oxford University Press. 1964.

Newmark, Joseph; Lake, Frances. *Mathematics as a Second Language.* New York: Addison-Wesley. 1982.

Ni, Hua Ching. *The Book of Changes and the Unchanging Truth.* Los Angeles: College of Tao & Traditional Chinese Healing. 1983.

Pagels, Heinz. *Perfect Symmetry.* New York: Simon & Schuster. 1983.

Peat, F. David. *Synchronicity: Bridge Between Mind and Matter.* New York: Bantam. 1987.

Peitgen, H.-O., & Richter, P.H. *The Beauty of Fractals.* Berlin: Springer-Verlag. 1986.

Prigogine, Ilya. *From Being to Becoming: Time and Complexity in the Physical Sciences.* San Francisco: W.H. Freeman & Company. 1980
Order Out of Chaos: Man's New Dialogue With Nature. With Isabelle Stengers. New York: Bantam Books. 1984.

Rothenberg, Albert. "The Emerging Goddess." *The Creative Process in Art, Science, and Other Fields.* Chicago: The University of Chicago Press. 1979.

Salisbury, Harrison. "A View From Mount Lu: Shedding 'A Little Blood.'" Paris: *International Herald Tribune.* June 15, 1989.

Schipper, Kristofer. Wang Hsiu-huei. "Progressive and Regressive Time Cycles in Taoist Ritual" in *Time, Science, and Society in China and the West; The Study of Time,* Vol. V. J. T. Fraser, N. Lawrence, & F. C. Haber, editors. Amherst: University of Massachusetts Press. 1986.

Schoenholtz, Larry. *New Directions in the I Ching.* Secaucus, New Jersey: University Books. 1975.

Schonberger, Martin. *The I Ching and the Genetic Code: the Hidden Key to Life.* New York: ASI Publishers. 1980.

Schrodinger, Erwin. *What Is Life?* England: Cambridge University Press. 1944.

Sessions, Roger. *Wisdom's Way: The Christian I Ching.* The Christian I Ching Society of Cedar Park, Texas. 2015.

Sheldrake, Rupert. A New Science of Life. Los Angeles: Tarcher. 1981.

Shubnikov, A.V., & Koptsik, V.A.; trans. editor, Harker, David. *Symmetry in Science and Art.* New York: Plenum Press, 1974.

Stent, Gunther. *The Coming of the Golden Age.* New York. Natural History Press. 1969.
The Double Helix, ed. by Gunther Stent. Norton Critical Edition. New York: W.W. Norton & Co. 1980.

Sung, Z D. *The Symbols of Yi King.* Shanghai: The China Modern Education Co. 1934.

Suzuki, David; Griffiths, Anthony; Miller, Jeffrey; Lewontin, Richard. *An Introduction to Genetic Analysis.* New York: W.H.Freeman. 1986.

Swami Ajaya. *Psychotherapy East and West: A Unifying Paradigm.* Honesdale, Penn: Himalayan Institute of Yoga Science. 1983.

Tien-Yien Li and Yorke, James A. "Period Three Implies Chaos." *American Mathematical Monthly,* Vol. 82. December, 1975.

Toffler, Alvin. *Future Shock.* New York: Bantam. 1971.

Vasavada, Arwind, with Spiegelman, J. Marvin. *Hinduism and Jungian Psychology.* Phoenix, Arizona: Falcon Press. 1987

Von Franz, Marie-Louise. "Dialog über den Menschen." Stuttgart: Klett Verlag. 1968.
Number and Time. Evanston: Northwestern University Press. 1974.

Waley, Arthur. *The Book of Songs*. Boston: Houghton Mifflin. 1937.

Three Ways of Thought in Ancient China. London: Allen & Unwin. 1939.

Watson, James. *Molecular Biology of the Gene*. New York: Benjamin, Inc. 1965.

Whincup, Greg. *Rediscovering the I Ching*. Garden City, N.Y.: Doubleday. 1986.

Wiggins, Stephen. *Global Bifurcations and Chaos: Analytical Methods*. New York-Berlin: Springer-Verlag. 1988

Wilhelm, Richard, trans.Cary F. Baynes. *The I Ching*. Princeton: Princeton University Press. 1950.

Wilhelm, Hellmut. Change: *Eight Lectures on the I Ching*. Trans. C.F. Baynes. Princeton: Princeton University Press. 1972.

Yan, Johnson F. *DNA and the I Ching: The Tao of Life*. Berkely, CA: North Atlantic Books. 1991.